Springer Proceedings in Physics

Volume 358

Indexed by Scopus

The series Springer Proceedings in Physics, founded in 1984, is devoted to timely reports of state-of-the-art developments in physics and related sciences. Typically based on material presented at conferences, workshops and similar scientific meetings, volumes published in this series will constitute a comprehensive up to date source of reference on a field or subfield of relevance in contemporary physics. Proposals must include the following:

- Name, place and date of the scientific meeting
- A link to the committees (local organization, international advisors etc.)
- Scientific description of the meeting
- List of invited/plenary speakers
- An estimate of the planned proceedings book parameters (number of pages/ articles, requested number of bulk copies, submission deadline).

Please contact:

For Americas and Europe: Dr. Zachary Evenson; zachary.evenson@springer.com
For Asia, Australia and New Zealand: Dr. Loyola DSilva; loyola.dsilva@springer.com

Paolo Cotta-Ramusino · Micah Lowenthal ·
Luciano Maiani · Barbara Gallo
Editors

Nuclear Weapon Risks in a Problematic Time

Proceedings of the XXIII Edoardo Amaldi Conference, Accademia Nazionale dei Lincei, Rome, Italy, 28–30 October 2024

Springer

Editors
Paolo Cotta-Ramusino
University of Milan
Milan, Italy

Luciano Maiani
Department of Physics
Sapienza University of Rome
Rome, Italy

Micah Lowenthal
CISAC
The National Academy of Sciences
Washington, DC, USA

Barbara Gallo
Istituto di Ricerche Internazionali Archivio Disarmo
Rome, Italy

ISSN 0930-8989 ISSN 1867-4941 (electronic)
Springer Proceedings in Physics
ISBN 978-3-032-29872-0 ISBN 978-3-032-29873-7 (eBook)
https://doi.org/10.1007/978-3-032-29873-7

This work was supported by Accademia Nazionale dei Lincei and Pugwash Conferences on Science and World Affairs.

The original submitted manuscript has been translated into English. The translation was done using artificial intelligence. A subsequent revision was performed by the author(s) to further refine the work and to ensure that the translation is appropriate concerning content and scientific correctness. It may, however, read stylistically different from a conventional translation.

This Springer imprint is published by the registered company Springer Nature Switzerland AG
The registered company address is: Gewerbestrasse 11, 6330 Cham, Switzerland

Preface

We live today in a world with regions riven by conflicts that have the potential to escalate to scales seen in the 20th century, or even far worse. Nearly 70 years ago, the first Pugwash conference established the model of scientists from different countries coming together to prevent, mitigate, and end conflicts that could result in nuclear war or proliferation. Those conferences continue to the present. Academies of sciences, which have special relationships with their governments, initiated their own activities that likewise have been carried forward to today.

In 1980, the US National Academy of Sciences established the Committee on International Security and Arms Control (CISAC) to open bilateral contacts with a parallel group simultaneously established at the Academy of Sciences of the Soviet Union. Among the CISAC founders was Wolfgang (Pief) Panofsky, eminent physicist, founder, and first director of the Stanford Linear Accelerator Center.

In January 1986, Panofsky led CISAC in a meeting on International Security and Arms Control that extended beyond the limitation of a direct link between the USA and the USSR. The possibility of establishing in Europe a group of scientists with a role analogous to CISAC was discussed. As a consequence, Edoardo Amaldi, in 1987, set up SICA, a working group on International Security and Arms Control, at the Accademia dei Lincei.

A first informal workshop was held in Rome, at Lincei Academy June 23–25, 1988 under the title International Security and Disarmament: The Role of the Scientific Academies. A second meeting followed in 1989.

On December 5, 1989, while preparing the third Meeting, Edoardo Amaldi, then President of Accademia Nazionale dei Lincei, unexpectedly died. The third meeting was held a few months later, in June 1990, under the Chair of Giorgio Salvini who had succeeded Amaldi at the Presidency of the Lincei. Participants at the third meeting decided to dedicate future meetings to the memory of Edoardo Amaldi, hence the name the Edoardo Amaldi Conferences.

From 1991 to 2011, the Edoardo Amaldi Conferences were regularly held at the Lincei and elsewhere in Europe, essentially as triangular meetings involving scientists from Europe, the United States, and the USSR, later the Russian Federation.

In 2015 and 2017, with the support of the Ministero degli Affari Esteri, the Edoardo Amaldi Conferences have been extended to important participation from Countries in the Middle East and Asia (China, Egypt, India, Israel, Iran, Japan, Pakistan).

SICA (Accademia dei Lincei) and CISAC (the US National Academy of Sciences) co-organized the 21st and 22nd conferences.

The 23rd Amaldi Conference was held in Rome, Accademia dei Lincei 28–30 October 2024, organized in collaboration with the US National Academy of Sciences (USA) and the Pugwash Organization. The Proceedings contain the texts of the sixteen talks given there.

The main focus of the conference was on the risk of possible nuclear conflicts generated by two active major conflicts, in Ukraine and in the Middle East, and the analysis of the possible consequences of local and global nuclear conflicts. Also discussed was the role of the scientific community to facilitate the restart of international dialogs on nuclear disarmament in the present situation, in particular in the East Asia environment. Finally, the role of artificial intelligence and space in the present conflicts, and the prospects of large-scale energy generation from nuclear fission and nuclear fusion were presented.

Talks were delivered by experts of different disciplines belonging to universities, international organizations, and governments from Europe, USA, Middle East, China, and Japan.

We appreciate the support of the Accademia Nazionale dei Lincei and of the US National Academy of Sciences, represented here by President Roberto Antonelli and by Dr. Micah Lowenthal, and of the Conference Board, Francesco Calogero, Carlo Schaerf, Francesco Lenci, Alessandro Pascolini, Alberto Quadrio Curzio, Raymond Jeanloz. We thank Barbara Gallo, Scientific Secretary of the Conference, and Mariella Di Donna, for the relations with the Press.

Rome, Italy

Luciano Maiani
Lincei Academy

Paolo Cotta-Ramusino
Pugwash Organisation

Micah Lowenthal
US National Academy of Sciences

Raymond Jeanloz
US National Academy of Sciences

Introduction

About the Conference

The Accademia Nazionale dei Lincei

Founded in 1603, the Accademia Nazionale dei Lincei is the oldest scientific academy in the world which included, among many other prestigious names, Galileo Galilei. The Accademia Nazionale dei Lincei is within the sphere of the Italian Ministry of Cultural Heritage and is considered the highest Italian cultural institution. The Accademia Nazionale dei Lincei is a scientific consultant to the President of the Italian Republic and has recently been granted his High Permanent Patronage. The Accademia Nazionale dei Lincei mission is to promote, coordinate, integrate and spread scientific knowledge in its highest expressions, in the frame of cultural unity and universality. To this end, the Accademia Nazionale dei Lincei organizes national and international conferences, meetings, and seminars.

The National Academy of Sciences

Established by an Act of Congress signed by President Abraham Lincoln in 1863, the National Academy of Sciences is charged with providing independent, objective advice to the nation on matters related to science and technology. The National Academy of Sciences is committed to furthering science in America, and its members are active contributors to the international scientific community. Nearly 500 members of the National Academy of Sciences have won Nobel Prizes, and the Proceedings of the National Academy of Sciences, founded in 1914, is today one of the premier international journals publishing the results of original research. The National Academy of Sciences also encourage education and research, recognize outstanding contributions to knowledge and increase public understanding in matters of science.

A Brief History of the Edoardo Amaldi Conferences

In 1980, the United States National Academy of Sciences established a Committee on International Security and Arms Control (CISAC). During the 1980s, with Wolfgang K. H. Panofsky as Chairman, one of CISAC's main purposes was to maintain bilateral contacts with an analogous group at the Academy of Sciences of the Soviet Union. In January 1986, CISAC organized in Washington, D.C. a meeting and about ten scientists from European countries participated. In this meeting questions connected to arms control were discussed, together with the possibility of establishing in Europe a group of scientists with a role analogous to that of CISAC. Discussions on international security and armament control were extended beyond the limitation of a direct link between the USA and the USSR.

Francesco Calogero and Carlo Schaerf took part in this meeting and then reported to Edoardo Amaldi, then Vice President of the Academia Nazionale dei Lincei. He was very much in favor of this initiative and set up a Working Group on International Security and Arms Control (SICA).

The first informal meeting was held in Rome at the Academy on June 23–25, 1988 and was entitled *Workshop on International Security and Disarmament: The Role of the Scientific Academies*. Participants from Belgium, Denmark, France, Federal Republic of Germany, Great Britain, the Netherlands, Sweden, and the USA were present at this meeting. On that occasion, it was decided to hold an international conference the following year and to invite participants from other European countries, including the Soviet Union.

The second meeting was entitled *International Security and Disarmament: The Role of the Scientific Academies* and was held in Rome in June 1989.

The third conference, entitled *International Conference on Security in Europe and the Transition Away from Confrontation Toward Cooperation*, was held again in Rome in June 1990. The title, the programme, and the specific items had already been established by Edoardo Amaldi (who was President of the Academy at the time) and the SICA group when he died unexpectedly on 5 December 1989. Professor Giorgio Salvini was elected to succeed him as President of the Academy and decided to continue the Amaldi's initiative on disarmament and arms control. The third meeting was devoted to problems of peace, environmental cooperation and measures of effective disarmament in the new international climate. It was at this meeting that the participants decided to dedicate all future conferences to the memory of Edoardo Amaldi, hence the name *Amaldi Conferences*. Since then the Amaldi Conferences continued on a yearly basis up until 2000, and later periodically.

Latest Conferences:

- XIX Edoardo Amaldi Conference on *International Cooperation for Enhancing Nuclear Safety, Security, Safeguards and Non-proliferation*, organized by the Accademia Nazionale dei Lincei and the European Commission—Joint Research Centre (Rome, 30–31 March 2015).
- XX Edoardo Amaldi Conference on *International Cooperation for Enhancing Nuclear Safety, Security, Safeguards and Non-proliferation—60 Years IAEA*

Atoms for Peace and Development—EURATOM Treaty, organized by the Accademia Nazionale dei Lincei and the European Commission—Joint Research Centre (Rome, 9–10 October 2017).
- XXI Edoardo Amaldi Conference on *International Cooperation for Enhancing Nuclear Safety, Security, Safeguards and Non-proliferation*, organized by the Accademia Nazionale dei Lincei and the National Academy of Sciences, USA (Rome, 7–8 October 2019).

Edoardo Amaldi (1908–1989)

Edoardo Amaldi was an Italian physicist who started his research in Nuclear Physics in collaboration with Enrico Fermi. He has contributed substantially to the current knowledge on artificial radioactivity induced by neutrons and properties of slow neutrons and gave first direct experimental evidence of the diffraction of fast neutrons by nuclei. After the Second World War, Amaldi was instrumental in the reconstruction of the post-Fermi School of Physics in Italy and was co-founder of the Italian National Institute for Nuclear Physics (INFN), of the European Organization for Nuclear Research (CERN) and of the European Space Research Organization (ESRO).

Amaldi supported the Working Group on International Security and Arms Control (SICA) from its inception, organizing the first conference on "International Security and Disarmament: The Role of the Scientific Academies" in Rome from 23 to 25 June 1988. The third conference on "Security in Europe and the Transition away from Confrontation toward Cooperation" was held again in Rome in June 1990. The title, the programme and the specific items had already been established by Edoardo Amaldi (President of the Academy at the time) and the SICA group when he died unexpectedly on December 5, 1989. Participants at this meeting decided to dedicate future meetings to the memory of Edoardo Amaldi, hence the name Amaldi Conferences. Since then the Amaldi Conferences have continued on a regular basis.

Speakers Who Attended the Conference Without the Written Contribution of their Academic Intervention

Alexey Arbatov, Russian Academy of Sciences, Moscow, Russia
Polycentric Nuclear Deterrence and Arms Control
Sergey Batsanov, Pugwash Conferences, Geneva, Switzerland
Increase of Nuclear Risks in the Times of Global Instability
Mariana Budjeryn, Harvard Belfer Center, Cambrige USA
Nuclear Risks in the Russian-Ukrainian War
Anton Khlopkov, CENESS, Center for Energy and Security Studies, Moscow, Russia
Nuclear Risks in Northen-Eastern Asia
Eldar Mamedov, Pugwash Council for Sciences and World Affairs
Iran and Escalation in the Middle East, Brussel, Belgium
Gotz Neuneck, University of Hamburg, Germany
Nuclear Weapons in Space—New Threats and Arms Control Solutions

Contents

Contributors

Hussain Al-Shahristani President of Pugwash, Baghdad, Iraq

Daniele Amoroso University of Cagliari, Cagliari, Italy

Nadezhda Arbatova Doctor, Professor of Political Sciences Head of the Department of European Political Sciences Primakov National Research Institute of World Economy and International Relations (IMEMO), Russian Academy of Sciences, Moscow, Russia

Fabrizio Battistelli Department of Social Sciences and Economics, University of Rome La Sapienza/Archivio Disarmo, Rome, Italy

Paolo Cotta-Ramusino Department of Physics, University of Milan, Milan, Italy

Persis S. Drell Stanford University, Stanford, CA, USA

Taghreed El-Khodary Palestinian Commentator and Media Expert, Brussels, Belgium

David Alexandre Ellwood British Pugwash, London, UK

Francesca Farruggia Department of Social Sciences and Economics, University of Rome La Sapienza, Rome, Italy

Francesco Lenci SICA International Security and Control Arms Committee, Accademia dei Lincei, Rome, Italy

Diego Mauri University of Palermo, Palermo, Italy

David Menashri Alliance Center for Iranian Studies, Tel Aviv University, Tel Aviv, Israel

Marc Mezard Bocconi University, Milan, Italy

Bruno Panella Department of Energy, Politecnico of Turin, Turin, Italy

Alessandro Pascolini University of Padova and SICA—International Security and Arms Control Committee, Rome, Italy

Francesco Pegoraro Physics Department, Pisa University, Pisa, Italy; Accademia Nazionale dei Lincei, Rome, Italy

Gian Piero Siroli Department of Physics and Astronomy, University of Bologna, INFN & CERN, Bologna, Italy

Tatsujiro Suzuki Research Center for Nuclear Weapons Abolition, Nagasaki University (RECNA), Nagasaki, Japan

Guglielmo Tamburrini University of Naples Federico II, Naples, Italy

Qizheng Zhang Center for Strategic Studies, China Academy of Engineering Physics, Beijing, China

Wuwen Zhao Center for Strategic Studies, China Academy of Engineering Physics, Beijing, China

Part I

Chapter 1
Nuclear Risks in Current Conflicts

Hussain Al-Shahristani

1.1 Keynote Lecture

It is a great pleasure to be back at the Amaldi conference. First, I would like to thank the Accademia Nazionale dei Lincei for their invitation and hospitality, and I hope you all enjoy your stay in the beautiful city of Rome.

I am asked to speak on nuclear risks in the current conflicts. The nuclear risks are not confined to the possible use of nuclear weapons in current world conflicts, but also the risks associated with the race to upgrade nuclear arsenals and resume nuclear testing and also risks of radioactive contaminants leakage from nuclear power plants in conflict zones. All these nuclear risks are real and should be of serious concern to us all.

Since the end of the cold war with the dissolution of the Soviet Union in Dec 1991, the world has not faced the risk of use of nuclear weapons as we are witnessing today. There are two ongoing wars in Ukraine and the ME, and a potential confrontation in east Asia where parties to the conflicts possess nuclear weapons. In the ongoing wars, there have been insinuations of possible use of tactical nuclear warheads.

But, let me first light on some dangerous developments concerning nuclear weapons. Even before these wars were waged, the major nuclear power states had started a new race to renovate and upgrade their nuclear arsenals.

In the United States, the Congressional Budget Office estimates that the United States is set to spend some $756 billion on nuclear weapons modernization programmes in the coming ten years. That is two and a half Manhattan projects every year.

The Russian Federation and the People's Republic of China are also actively upgrading their nuclear facilities and arsenals. There are no published figures on budget allocation for these programmes. Russia, the traditional nuclear peer to the

---

H. Al-Shahristani (✉)
President of Pugwash, Baghdad, Iraq
e-mail: hussain@alshahristani.com

P. Cotta-Ramusino et al. (eds.), *Nuclear Weapon Risks in a Problematic Time*, Springer Proceedings in Physics 358, https://doi.org/10.1007/978-3-032-29873-7_1

US, is already engaged in renovating its nuclear arsenal and developing a new "exotic" nuclear weapons and delivery missiles.

The nuclear modernization in the US and Russia, has now grown from just updating older systems, to the development of entirely new more potent nuclear weapons. Both countries are considering manufacturing what is referred to as "tactical" or "battlefield nuclear weapons". Such weapons are extremely dangerous because being smaller and less destructive, countries at war are tempted to use them. Low-yield nuclear weapons are not deterrent to the warring parties, quite the contrary, they may make nuclear use more likely.

The other nuclear country that we should be concerned about is PRC. China, as an emerging nuclear superpower, has accelerated its nuclear expansion. It is expected that it would field over 1000 operational warheads by 2030. They have completed silo construction and begun loading them with missiles. China is not a nuclear peer to the United States yet, but at the current rate of expansion, it could reach that status in less than a decade. Unlike Russia, China would be a peer to the United States not only in the nuclear field, but also in other economic and technology domains.

All the three countries are reopening and upgrading their nuclear test sites apparently to test the new nuclear warheads under development despite that they are signatories to the Comprehensive Nuclear-Test-Ban Treaty since 1996. Russia had even ratified the treaty, but last year it revoked its ratification. It is to be seen which of the three would breach the treaty first in the coming years.

The problem with such new nuclear weapons' development is that it is highly destabilizing and could lead to a new nuclear race. When any nation starts such a race, its nuclear rivals would surely react by upgrading their nuclear deterrence to offset any strategic advantage its rival might gain to maintain deterrent parity at the least. Or at worst, they may adopt an offensive nuclear doctrine to take out the new enemy nuclear silos and facilities before they are operational.

Of course, such an extreme scenario is highly unlikely given rational thinking of nuclear adversaries when there is a balanced deterrence as was the case during the cold war years when the Mutual Assured Destruction doctrine prevented nuclear superpowers from resorting to such means. But, when a nuclear power feels that its nuclear deterrence may not protect it against an enemy nuclear strike, do we have any assurance that it may not resort to a pre-emptive strike to take out the enemy offensive nuclear advantage.

What makes matters worse is that we will soon face a 3-body problem instead of the 2-body that existed during the previous cold war. And we all know how much more difficult it is to solve a 3-body problem compared to 2-body one.

Other sources of serious concern for potential nuclear warfare are the advancements in new disruptive technologies within the domains of hyper-sonics, anti-submarine warfare, weaponizing the space, missile defence and AI-enabled data processing. These fast-developing technologies are changing the role of nuclear weapons. These innovations are encouraging some policy makers and strategists to reconsider the foundational assumptions about deterrence and risk.

We hear with great alarm proposals to consider a pre-emptive launch of nuclear strike to disarm adversaries. The mere debate on the rationality of such doctrine by some policymakers is extremely alarming, particularly when it comes at a time when discussions on extension of the New START treaty is stalled. This treaty that limited the number of deployed strategic nuclear warheads in the US and Russia to 1550 will expire in Feb. 2026.

Instead of engaging in discussions to renew this treaty or replacing it with a better one, we notice signals that this important treaty may not be renewed, and the superpowers could initiate a new nuclear race with more sophisticated warheads.

President Putin addressing the Russian Federal Assembly in Feb 2023 announced that Russia would suspend its participation in New START but is not withdrawing from it. However, Russia clarified that it would continue to abide by the numerical limits in the treaty but would discontinue the notification process.

War between nuclear armed states is not impossible. Nuclear warfare is not an abstract danger, but a real one. Bearing in mind regional and global tensions, several plausible scenarios exist today in different parts of the world for nuclear exchange.

Against such troubling background, let us focus on the current world conflicts: the Ukraine war, and the conflicts in the ME.

First let us consider the potential use of nuclear weapons in the Ukraine war.

The war in Ukraine involves one nuclear-weapon state directly, Russia, and three nuclear-weapon states indirectly, the US, France and UK. The later are considering material assistance to Ukraine to attack military sites inside Russia. Should the NATO allow long range missile attacks against strategic military sites in Russia, that could prompt a Russian response with tactical nuclear warheads.

President Putin has declared two years ago in Sept 2022 that Russia "will use all the means at our disposal in order to defend the country's territory". This was widely understood as using nuclear weapons. This announcement was followed six months later by moving Russian tactical nuclear weapons to the Republic of Belarus.

More recently, he warned that supplying long-range weapons to Ukraine to hit deep inside Russia would result in a nuclear conflict. He stated that "we also have the weapons that can strike targets on their territory, and what they are now suggesting and scaring the world with, all that raises the real threat of a nuclear conflict that will mean the destruction of our civilization".

On the other side, President Zelensky has threatened to seek nuclear weapons unless Ukraine is given NATO membership. He told European Union leaders during a speech in Brussels: "Either Ukraine will have nuclear weapons, which will serve as protection, or it must be part of some kind of alliance. Apart from NATO, we do not know of such an effective alliance".

These threats should not be dismissed or ignored. Yes, for the time being, it is quite unlikely that there will be a nuclear exchange in Europe as a result of the ongoing war in Ukraine, but the possible detonation of even a small tactical nuclear warhead is extremely alarming. Such detonation would have devastating and compounding effects, and carry a high risk of triggering an escalation to a limited or all-out nuclear war.

The nuclear risk in the Ukraine conflict is not limited to direct use of nuclear weapons. The Ukrainian nuclear power stations have been caught in the crossfire on a number of occasions. The largest nuclear power plant in Europe, the Zaporizhzhia

Nuclear Power Plant that the Russian forces had seized in March 2022 is right at the battle frontline. This nuclear power plant is now badly damaged, and it could be accidentally hit by the exchange of fire by the combatants causing release of radioactive materials.

Some of us here have vivid memory of the Chernobyl nuclear power accident in northern Ukraine in 1986. Then radioactive nuclides from Chernobyl contaminated large areas of Central Europe.

According to the International Atomic Energy Agency report, "The situation in Ukraine is unprecedented. It is the first time a military conflict has occurred amid the facilities of a large, established nuclear power". Grossi, Director General of the IAEA, reported to the annual meeting of Member States last month that Zaporizhzhia plant still suffers from "Regular explosions, drone attacks, gunfire and repeated interruptions of external power supply, among other challenges". The risk of a nuclear accident at Zaporizhzhia plant is real.

What complicates the matter further is that the Russia's Kursk nuclear power plant also came under threat when Ukrainian forces advanced across the border. The Kursk nuclear power plant is one of the three largest in Russia. The threat to these nuclear facilities underscores how both Russia and Ukraine view nuclear power plants as strategic targets and their skirmishes near these plants could lead to a disastrous radioactive contamination to the area and beyond.

The other conflict that has a potential nuclear risk is the war in the Middle East.

The ME is one of the most volatile regions in the world. Wars and conflicts there do not only destabilise the region, but also affect peace and security further away. On several occasions in the past five decades, parties to military confrontations in the region either used or were on the verge of using weapons of mass destruction. Chemical weapons were used by Saddam Hussein in the Iran-Iraq war, and they were used in the Syrian civil war.

Currently, there are two conflicts that undermine the stability of the region: the war in Gaza and Lebanon, and presence of nuclear warheads and nuclear activities that could lead to nuclear race and confrontation.

I will touch on these two issues briefly. First let us consider the nuclear risks. Israel possesses nuclear weapons. Iran is pursuing a uranium enrichment programme to produce 60% enriched uranium. After years of laborious negotiations, the P5 + 1 reached the JCPOA agreement with Iran in 2015, which the US President Trump unilaterally abrogated. This has put the region back in unchartered nuclear waters.

It is unfortunate that Dr. Ali Salehi, one of the architects of JCPOA, who was invited to this conference could not travel due to the current security situation in the ME. He could have elaborated on the current status of the agreement and if there is anything that can be done to return to the JCPOA or reach to a new similar agreement.

Iran has repeatedly stated that it does not intend to produce nuclear warheads, and nuclear weapons are not part of their defence doctrine. And the new government in Tehran has expressed interest to resume talks on implementation of the JCPOA, but the USA is not expected to respond to this gesture till the new administration takes charge.

We hope the new administrations in Washington and Tehran would seriously consider salvaging that agreement or attempt to replace it with an updated one. Of course, the three European countries that are party to the agreement and the EU can play a constructive role in reaching such an outcome.

On the positive side, there are some encouraging developments in the region. The relationship between the Arab Gulf States and Iran has improved, and there are proposals for cooperation on peaceful applications of nuclear energy and reginal nuclear security arrangements.

However, the ultimate guarantee for security in the ME is the elimination of nuclear and other weapons of mass destruction from the region. The UNGA had adopted a resolution to establish a nuclear weapons free zone (NWFZ) in the ME fifty years ago in 1974. Since then, the GA reaffirmed its resolution almost every year. All countries in the region expressed willingness to join such a treaty except Israel.

Nuclear weapons do not prevent wars or provide security. Israel's possession of nuclear warheads has not prevented four major wars with its neighbours and did not contribute to bringing an end to the war in Gaza that has now spread to the West Bank and Lebanon. Similarly, the use of chemical weapons by Saddam Hussein did not end his war with Iran, and possession of such weapons by Moamar Qaddafi did not save the Libyan dictator.

The second issue in the ME is the Palestinian question and more urgently the war in Gaza that has been going on for more than a year now, and the war in Lebanon that flared up earlier this month. More than 160 thousand civilians so far have been killed or severely wounded in Gaza, the majority of whom are women and children. The population in the strip is near starvation and totally depend on UNRWA for basic necessities. Even this life-saving channel has now been curtailed by cutting funding by the US and some European countries.

The war in Gaza has reached a genocidal proportion. It must stop to save the remaining of the Palestinian people. Otherwise, in addition to its human cost, it will push the ME to roll down to a regional war with catastrophic consequences not only to the region but to the world at large.

This war has now spread to Lebanon and could drag in other countries. Iran has fired a barrage of hypersonic missiles at Israel in response to the killing of Ismail Haniyeh in Tehran and Sayed Hassan Nasrullah in Beirut. Israel has reciprocated with a limited attack last Saturday, and Iran said it will respond proportionally. This calculated exchange is reassuring. But it should not be ruled out that miscalculations could lead to a spiral mutual escalation ending with uncontrolled regional war.

The international community, particularly the US and European countries should act directly and through the UN to end this war and prevent any further escalation. The Security Council should be pressed to adopt a resolution for immediate ceasefire in Gaza, and the permanent members of the council should not hold up such a resolution by their veto power. Pending cessation of hostilities, all countries should refrain from sending offensive weapons to the region.

We in the Pugwash Conferences on Science and World Affairs issued an appeal to all parties involved in the conflict to abstain from targeting nuclear and/or energy infrastructure in view of the devastating consequences that this could unleash not only on the region but the world at large.

We called on all parties to the conflict to abide by the International Humanitarian Law and Article 56 of the Additional Protocol I to the Geneva Conventions that specifically prohibits attacks on nuclear installations, recognizing that such actions pose an indiscriminate threat to the civilian populations. The deliberate targeting of nuclear and energy facilities also violates the Environmental Protection Provisions under IHL, which demand that parties to a conflict avoid actions that cause widespread, long-term, and severe damage to the environment.

The world peace and security cannot be maintained without upholding international laws and conventions. Any violations or deviations from adherence to them should be unequivocally condemned whenever and wherever they occur.

The humanitarian disaster in Gaza cannot be averted without enabling UNRWA to carry out its mandate of providing humanitarian assistance to Gazans who have been the victims of this war. Instead of pledging more funds to this vital UN agency to save the remaining population in Gaza from starvation, the major Western donors have suspended their contributions.

Israel has also expanded its war into Lebanon. This dangerous development can draw in other actors, and the war may easily engulf the whole region. The consequences of such a disastrous development would be devastating. Other countries, including nuclear countries, may be drawn in.

Risks of Nuclear strike in the war in the ME although very unlikely, but it should not be ignored. Israel may be tempted to use its nuclear weapons if it sees itself overwhelmed as it had considered this option in the 1973 war with Egypt. There have been occasional calls by some politicians in Israel to use their nuclear weapons to end the war in Gaza. Even a senior US senator has suggested it. Such an action will most likely force Iran to reconsider its defense doctrine and dash to develop its nuclear weapons.

I am not predicting such a doomsday scenario. I am merely drawing the attention of responsible scientists like yourselves to the dangers of letting this war to continue and expand. We must act now by pressuring our governments to act unilaterally or multilaterally to force the combatants to end this war immediately and seek a peaceful resolution of the conflict.

Chapter 2
Reducing Risks of Nuclear Weapons Use in Northeast Asia

Tatsujiro Suzuki

Abstract The risks of nuclear weapons are said to be the highest since the end of Cold War, primarily due to increasing tensions among nuclear armed states. In Northeast Asia. There are four nuclear armed states (US, China, Russia, DPRK) and the risks of nuclear weapons use is also increasing in this region. This paper is based on a three-year project to reduce risks of nuclear weapons use in Northeast Asia, by examining plausible cases of nuclear weapons use, quantification of consequences of nuclear weapons uses on selected five cases and draw policy implications for recommended actions to reduce the risks of nuclear weapons use in Northeast Asia. The most important policy implications are: (1) Unintended nuclear weapons use can be triggered by miss-communication, lack of mutual understanding, accident and human errors, (2) Once nuclear weapons are used, it is very difficult to foresee what will follow, and there is always a possibility that it could lead to a global nuclear war, (3) Use of nuclear weapons could result in unparallel humanitarian consequences, (4) First nuclear weapons use are likely to be "tactical nuclear weapons" and the first targets are likely to be military targets in "nuclear umbrella" states. The study concluded that policies oriented towards increased, stronger, or enhanced deterrence are making the region less secure. To make ambitious cooperative measures aimed at reducing the threat of nuclear weapons must be taken by the leaders in the region.

2.1 Introduction

It is now 89 Seconds to Midnight". Bulletin of Atomic Scientists' "Dooms Day Clock" [11] is now closer than ever to the human catastrophe. Bulletin cites the increasing nuclear risk as the primary reason for such global risk.

Global stockpile of nuclear warheads is now increasing. According to the latest estimate by Research Center for Nuclear Weapons Abolition, Nagasaki University (RECNA), total number of nuclear warheads is now 12, 340 (as of June 2025) compared to 12, 120 in 2024. Furthermore, total number of "deployable nuclear

T. Suzuki (✉)
Research Center for Nuclear Weapons Abolition, Nagasaki University (RECNA), Nagasaki, Japan
e-mail: suzukitatsu@nagasaki-u.ac.jp

P. Cotta-Ramusino et al. (eds.), *Nuclear Weapon Risks in a Problematic Time*, Springer Proceedings in Physics 358, https://doi.org/10.1007/978-3-032-29873-7_2

warheads" (i.e. total of deployed warheads and reserved warheads) is now 9615 compared to 9583 in 2024. In fact, the latter number has been increasing steadily since 2018 from 9251 to 9583, i.e. increase of 332 warheads [10].

Regional conflicts, such as those in Ukraine, Gaza and South Asia, may increase the possibility of nuclear weapons use for the first time since 1945. It is true that nuclear weapons have never been used since the droppings of atomic bombs over Hiroshima and Nagasaki in 1945. But there is no evidence that "nuclear deterrence" is the primary reason for such "non-use" of the nuclear weapons. Instead, it can be said that there has been a growing international norm that nuclear weapons should never be used again. This norm has become known as "the nuclear taboo" [12]. In fact, the Nobel Peace Prize committee awarded 2024 Nobel Peace Prize to Nihon Hidankyo (Japan Confederation of A-and H-bomb Suffers Organization) for their contribution to spreading the norm of nuclear taboo, but adding that "this taboo against the use of nuclear weapons is under pressure" [6].

Now we are facing a critical time whether we can maintain such nuclear taboo.

2.2 Increasing Nuclear Risks in Northeast Asia

Northeast Asia is one of the regions that nuclear risks are considered "high", since there are four nuclear armed states (US, Russia, China and DPRK) and there are increasing tensions among those countries.

US-Russia nuclear disarmament efforts have been stalled since the first Trump Administration formally announced its intention to withdraw from the Intermediate-Range Nuclear Forces (INF) Treaty in 2018. In response, Russian President Vladimir Putin also announced that Russia will be also officially suspending its treaty obligations as well. Since Russian invasion of Ukraine in 2022, the relationship between Russia and the US has grown worse, and there has not been any negotiation to extend the New START (Strategic Arms Reduction Treaty) which will expire in February 2026. If the Treaty is not extended, or no new Treaty replacing the New START, there will be no bi-lateral disarmament treaty between the two largest nuclear armed states.

China is the only nuclear weapon states who declare "no-first-use" policy but is currently rapidly expanding nuclear forces. The number of nuclear warheads has increased dramatically from 250 in 2015 to 350 in 2030 and now 600 in 2025. It is not clear why China has been rapidly increasing its nuclear arsenals [1], but some experts observe that China is responding to US modernization program of nuclear weapons [2]. US Department of Defense estimated that the number of Chinese nuclear weapons might increase more than 1000 by 2030 [13].

DPRK is also expanding its nuclear forces and missile programs steadily. More importantly, their nuclear policy has been transforming into the one under which they are ready to use nuclear weapons first. In 2022, the new "Nuclear Forces Policy Law" clarifies the conditions of nuclear weapons use which are: (1) attacked by nuclear weapons or other weapons of mass destruction or attacked in a scale similar to the above ones, (2) hostile attacks (either nuclear or non-nuclear) against national leader or nuclear strategic command organizations, (3) fatal attack (either nuclear or non-nuclear) against critical facilities, (4) to prevent escalation of conventional war and it is considered essential to use nuclear weapons for controlling the warfare, (5) other conditions under which very existence of nation or catastrophic national crisis to endanger citizens of our nation [3]. Most recently, in June 2024, DPRK and Russia announced that it concluded "Comprehensive Strategic Partnership Treaty" between them. It means that if either one of them is involved in a war, the other will assist its partner with military and other aid [4].

In response to both Chinese and DPRK threats, Republic of Korea (ROK), Japan and the United States agreed to strengthen their "extended nuclear deterrence" arrangement. In 2023, the leaders of the three nations met in Camp David and their joint statement said, "And we will enhance strategic coordination between the U.S.-Japan and U.S.-ROK alliances and bring our trilateral security cooperation to new heights" [5].

2.3 RECNA's Project on Reducing Risks of Nuclear Weapons Use in Northeast Asia

2.3.1 Plausible Cases of Nuclear Weapons Use

Given the increasing tensions in the region where four nuclear-armed states exist, Research Center for Nuclear Weapons Abolition, Nagasaki University (RECNA), in collaboration with Nautilus Institute for Security and Sustainability and Asia Pacific Leadership Network (APLN), conducted a three-year study entitled, "Reducing the risk of Nuclear Weapons Use in Northeast Asia" from April 2021 to March 2024. The objectives of this project are: (1) to understand the risk of nuclear weapons as well as conditions leading to the first use of nuclear weapons, (2) to develop policy recommendations to minimize such risks, (3) to prevent any use of nuclear weapons in the region.

At first year, the study developed 25 plausible cases of nuclear weapons use in the region [7]. The project convened a virtual workshop in October 2021, inviting experts on nuclear strategy, security, domestic politics of nations in the region, and nuclear disarmament and non-proliferation, from the US, ROK, Japan, Russia, Australia and China. The project focused on "triggering events" leading to first use of nuclear weapons, and "plausibility" based on reliable evidence is one important criterion to develop cases. The study put priority in policy implications and eventually 25 cases were developed.[1] 25 cases were classified depending on the "Intent" (whether it was

[1] After Russian invasion of Ukraine in 2022, the study added another five cases for the second-year project.

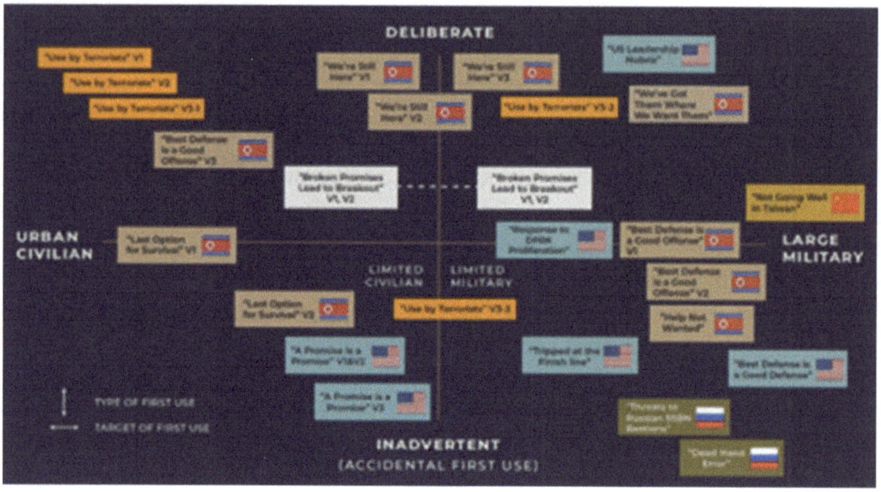

Fig. 2.1 Plot of use cases versus intent and type of target

"intended" or "inadvertent") and the "target" ("military target" or "urban/civilian target") (Fig. 2.1).

Figure 2.1 shows the variation of all 25 cases.

Based on the 25 cases, the study drew policy implications as follows:

- About half of the cases involve "un-intended (inadvertent)" use of nuclear weapons. Such "un-intended" use can be triggered by miss-communications, lack of mutual understanding, accidents and human errors etc.
- Once nuclear weapons are used, it is very difficult to foresee what will follow. Diplomacy could end the conflict, but developments after the first use are unpredictable. Regardless of intentions by policy/military leaders, it is possible that the first use of nuclear weapons could develop into uncontrollable nuclear exchange.
- Personalities and behaviors of leaders under pressure (crisis situation) who control the "button" could be critically important.
- Diversity of nuclear weapons and nuclear strategies makes it more difficult to deal with nuclear weapons under crisis situation.

2.3.2 Consequences of Nuclear Weapons Use: 5 Cases

The study shifted its focus on the consequences of nuclear weapons use in Northeast Asia in the second year. The study selected five cases out of 30 cases developed in previous year, considering the diverse consequences of nuclear weapons use. Here is the brief description of five cases (see Table 2.1) [8].

1. "We are still Here": The DPRK, facing internal and external economic pressures, undertakes what be thought as a "demonstration" nuclear attack over a small ROK

Table 2.1 Summary of five selected cases

Evaluated Use Case	First User	Responding User(s)	Weapon Sizes	Total Detonations
#1: "We're Still Here" Variant 1	DPRK	United States	10 kT (fission), 8 kT (2-stage H-bombs)	**3**
#2: "US Leadership Hubris"	United States	DPRK, China	20 and 10 kT (fission), 8, 50, 200, and 300 kT (2-stage H-bombs)	**18**
#3: "Terrorist" Variant 1	Terrorist	[None]	10 kT (fission)	**1**
#4: "Conflict from Ukraine Spreads East"	Russia	United States	2-stage H-bombs, 150, 200 kT, and 8 kT	**8**
#5: "Not Going Well in Taiwan"	China	United States, China	8, 50, 250, and 300 kT (2-stage H-bombs)	**24**

community for the main purpose of driving the US, the ROK and the international community to the bargaining table. In response, the US first attacked DPRK with conventional forces and used low-yield nuclear weapons to attack buried targets concealing ICBM and other nuclear weapons in DPRK. In total three nuclear weapons are used.

2. "US Leadership Hubris": Overconfident the US president and his advisors made a decision to attack on the DPRK's nuclear and missile systems using both conventional and nuclear weapons. But the mission accomplished only partially, the DPRK responded to attack US military bases and industrial targets both in ROK and in Japan with nuclear weapons. Additional attack by the US with nuclear weapons led China to respond, by attacking US bases in the region. A total of 18 nuclear weapons are used.
3. "Terrorist attack" (Variant 1): A terrorist group (unknown) stages an attack to gain worldwide attention by detonating nuclear device in metropolitan area of Tokyo. Nuclear forensic analyses points to the DPRK as a source of nuclear material but is not strong enough to conclude, and the DPRK denies any role in attack. Only one small nuclear explosive device was used.
4. "Conflict from Ukraine spreads to East": Given the increased threats not only from China/DPRK but also from Russia, Japan and the ROK urged the US to bring more nuclear weapons in the region. The alarmed Russia decided to increase the level of nuclear weapon system to higher alert status. Ultimately, fearing a US attack, Russia decided to attack US military bases (naval bases and military ships and bases in Japan) in the region with nuclear weapons. The US responds to attack Russian Far East nuclear bases with nuclear weapons. A total of 8 nuclear weapons are used.
5. "Not Going well in Taiwan": In order to distract public in China, Chinese leaders decided to attack Taiwan with conventional forces. The US rapidly involved in the conflict and China suffered significant setback. Worried about further setbacks and to prevent escalation, China first threatens to use nuclear weapons, despite of its "no-first-use policy." With little effect of such threats, China attacks US

bases in the region with nuclear weapons. In response, the US attacked Chinese nuclear and missile bases in China. A total of 24 nuclear weapons are used.

The results of estimated deaths for each case are summarized in Table 2.2.

In each case, computer simulations showing the spread of radioactive fallout was developed by the study team. Figure 2.2 shows one of such simulation results for Case 5.

Given the results of estimated deaths and spread of radioactive fallout, the study drew policy implications as follows.

- Use of nuclear weapons could result in unparallel humanitarian consequences compared with conventional weapons both in quantity and quality. Even an attack

Table 2.2 Summary of the estimated deaths for each case

Estimated Likely Deaths	Prompt (days to weeks)	Short-Term (weeks to months)	Additional Impact: Firestorms	Total Fatalities within 0.5 psi Zone (Total Population, % Lethality)	High Radiation Dose (Fallout) (short-term deaths)	Radiation-induced Cancer (long-term deaths)
Use Case 1 Airburst: 1, Surface-burst: 2	5,500	5,600	Firestorm Unlikely	**11,000** (41,000, 27%)	Low Fallout	16,000 - 36,000
Use Case 2 Airburst: 11, Surface-burst: 7	1,100,000	810,000	170,000	**2,100,000** (6,200,000, 33%)	11,000 - 1,200,000	480,000 - 920,000
Use Case 3 Surface-burst: 1	82,000	140,000	Small Centralized Firestorm	**220,000** (890,000, 25%)	0 - 1,600,000	410,000 - 560,000
Use Case 4 Airburst: 8	170,000	100,000	15,000	**290,000** (800,000, 36%)	Low Fallout	14,000 - 85,000
Use Case 5 Airburst: 16, Surface-burst: 8	1,500,000	930,000	190,000	**2,600,000** (7,600,000, 35%)	400 - 19,000	96,000 - 830,000

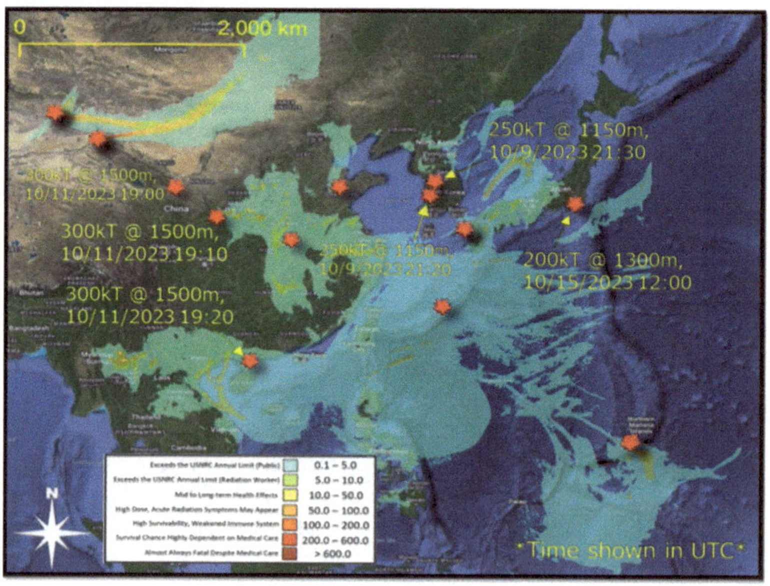

Fig. 2.2 Simulation results of radioactive cloud for case 5

against remote areas with small nuclear devices could result in thousands of casualties. It is also now clear that radioactive fallout could diffuse widely, affecting the regions outside the conflict areas.

- Once nuclear weapons are used, there is always a risk that it could lead to global nuclear war in a very short time (within hours or days). There are many pathways leading to the first use of nuclear weapons (with or without intentions).
- First nuclear weapons used are likely to be "tactical (non-strategic) nuclear weapons" and the first targets are likely to be military targets in "nuclear umbrella" states, such as those in ROK and Japan.

2.4 Recommended Actions to Reduce Nuclear Risks

Based on the results of the first year and the second-year research shown above, the study made 23 practical policy recommendations to reduce nuclear risks in the region [9]. Most of the recommended policies are the ones once considered to be implemented or once introduced in the past. The recommended actions are divided into three steps in order of "practicality" or "easiness" of implementation. Some selected recommendations are the following.

1. Warming Actions (Rhetorical and diplomatic gestures aimed at alleviating tension in the security environment) (three examples of six actions)
    - Rescope extended deterrence dialogues between Japan, ROK and US for risk reduction.
    - Declare mutual co-existence with China and DPRK, intention to end the Korean War and start strategic dialogue with DPRK by the US
    - A "No leadership Assassinations" pledge.
2. Ripening Actions (Individual Restraint) (three out of ten actions)
    - A "no nuclear deployment" in NE Asia by the US
    - Support the restricting First Use of Nuclear Weapons Act (of US Congress)
    - Pause and investigate permanent halt of development and deployment of ground-based intermediate-range missiles.
3. Reciprocal Transformations (Bilateral or Multilateral Initiatives) (three examples out of seven actions)
    - Multilateralize a moratorium on anti-satellite testing
    - A ban on low-yield "tactical" nuclear weapons
    - A "nuclear-free seas" initiative with North Korea.

The leaders of each country in the region could select any one or combination of above recommended actions depending on preferences of practicality reflecting on the current severe security environment. Any combination of such actions is expected to reduce the risk of nuclear weapons considerably.

2.5 Conclusions

Northeast Asia is barreling toward nuclear precarity. Policies oriented towards increased, stronger, or enhanced deterrence are making the region less secure. To make ambitious cooperative measures aimed at bridling the threat of nuclear weapons politically feasible, governments must warm and ripen the regional security environment so that leaders are able to embrace a less destructive path.

The declaration, "Let Nagasaki be the Last!" is an ambitious demand, matched by our ambitious proposals. We call on leaders from the United States and Northeast Asia to help the world ensure that Nagasaki will indeed be the last.

References

1. F.S. Cunningham, The unknowns about China's nuclear modernization program. Arms Control Assoc. (2023). https://www.armscontrol.org/act/2023-06/features/unknowns-about-chinas-nuclear-modernization-program
2. M. Fravel, H. Taylor, S. Henrik, M.L. Trøan, China's misunderstood nuclear expansion: how U.S. strategy is fueling Beijing's growing arsenal. Foreign Aff. (2023). https://www.foreignaffairs.com/china/chinas-misunderstood-nuclear-expansion?check_logged_in=1/
3. Korean Central News Agency, Law on DPRK's policy on nuclear forces promulgated (2022). http://kcna.co.jp/item/2022/202209/news09/20220909-02ee.html
4. Korean Central News Agency, *DPRK-Russia Treaty on Comprehensive Strategic Partnership* (2024). http://kcna.kp/en/article/q/6a4ae9a744af8ecdfa6678c5f1eda29c.kcmsf
5. Ministry of Foreign Affairs, Japan, The spirit of camp David: joint statement of Japan, the Republic of Korea, and the United States (2023). https://www.mofa.go.jp/files/100541826.pdf
6. Nobel Peace Prize Committee, *Nobel Peace Prize 2024 Press Release* (2024). https://www.nobelprize.org/prizes/peace/2024/press-release/
7. Research Center for Nuclear Weapons Abolition, Nagasaki University (RECNA), *Possible Nuclear Use Cases in Northeast Asia: Implications for Reducing Nuclear Risk* (2022). https://www.recna.nagasaki-u.ac.jp/recna/bd/files/Year_1_NU-NEA_Report_E_220128-1.pdf
8. Research Center for Nuclear Weapons Abolition, Nagasaki University (RECNA), *Humanitarian Impacts of Nuclear Weapons Use in Northeast Asia: Implications for Reducing Nuclear Risk* (2023). https://www.recna.nagasaki-u.ac.jp/recna/bd/files/Year_2_NU-NEA_Report_E_2303
9. Research Center for Nuclear Weapons Abolition, Nagasaki University (RECNA), *What should be done?—Practical Policies to Prevent Nuclear Catastrophe* (2024). https://www.recna.nagasaki-u.ac.jp/recna/bd/files/Year_3_NU-NEA_Report_E_2403.pdf
10. Research Center for Nuclear Weapons Abolition, Nagasaki University (RECNA), World's nuclear warheads data (2025). https://www.recna.nagasaki-u.ac.jp/recna/en-topics/48377
11. Science and Security Board, Bulletin of Atomic Scientists, Doomsday clock statement (2025). https://thebulletin.org/doomsday-clock/2025-statement/
12. N. Tannenwald, *The Nuclear Taboo: The United States and the Nonuse of Nuclear Weapons Since 1945* (Cambridge University Press, 2007)
13. US Office of the Secretary of Defense, *Annual Report to Congress: Military and Security Developments Involving the People's Republic of China 2023* (2023). https://media.defense.gov/2023/Oct/19/2003323409/-1/-1/1/2023-MILITARY-AND-SECURITY-DEVELOPMENTS-INVOLVING-THE-PEOPLES-REPUBLIC-OF-CHINA.PDF

Chapter 3
Controlling AI? A Technological Revolution Waiting for a Theory

Marc Mezard

Abstract This paper explores the rapid evolution of Artificial Intelligence (AI), highlighting its historical landmarks, technological advancements, and the growing concerns regarding its interpretability, ethical considerations, and societal impacts. The discussion emphasizes the challenges posed by the opaque nature of deep learning models, their implications for security and decision-making, and the risks associated with autonomous AI-driven systems. Additionally, we discuss the necessity of a theoretical foundation for AI to ensure its responsible and explainable deployment.

3.1 Introduction

Thanks to the development of deep neural networks trained with machine learning, artificial Intelligence has undergone a technological revolution in recent decades. While AI systems have demonstrated remarkable capabilities in various domains, and in spite of the fact that one knows all the parameters of a trained neural network, the underlying mechanisms of how it reaches its results remain largely unexplained. This paper summarizes the trajectory of AI development, insisting on its theoretical gaps, and the pressing need for a deeper understanding.

The rapid acceleration of AI-driven applications raises fundamental questions about control, explainability, and trust. From chatbots to autonomous decision-making systems, the integration of AI into everyday life necessitates an urgent exploration of its implications.

M. Mezard (✉)
Bocconi University, Milan, Italy
e-mail: marc.mezard@unibocconi.it

P. Cotta-Ramusino et al. (eds.), *Nuclear Weapon Risks in a Problematic Time*, Springer Proceedings in Physics 358, https://doi.org/10.1007/978-3-032-29873-7_3

3.2 Historical Milestone in the Development of AI

The development of artificial intelligence has been punctuated by several pivotal breakthroughs. These milestones not only reflect technological advancement but also the shifting understanding of learning, perception, and reasoning in artificial systems.

1958: The Perceptron

The Perceptron, introduced by Frank Rosenblatt, was one of the earliest models for binary classification inspired by the structure of the brain. It formalized a simple neural unit that computes a weighted sum of inputs and applies a threshold. Despite its limitations—notably its inability to learn non-linearly separable functions, as pointed out by Minsky and Papert—the Perceptron laid the conceptual foundation and the basic building block for later neural network architectures.

1980s–1990s: Neural Network Renaissance

The 1980s and 1990s witnessed a resurgence in neural network research, driven by key algorithmic and architectural advances. The backpropagation algorithm, popularized by Rumelhart, Hinton, and Williams, enabled efficient training of multilayer networks. At the same time, convolutional neural networks (CNNs), introduced by LeCun and colleagues, demonstrated remarkable performance in image recognition tasks. Meanwhile, Boltzmann machines provided a probabilistic framework for learning complex representations, laying the groundwork for generative models. An influential contribution from this period was the network, introduced by John Hopfield, which modeled associative memory using recurrent neural dynamics. The Hopfield model bridged ideas from physics, particularly spin glass theory, and neuroscience, capturing how neural systems might store and retrieve information. These developments helped forge deeper ties between AI and neuroscience, culminating in the emergence of computational neuroscience as a field devoted to the mathematical modeling of neural processes and cognition.

2012: AlexNet and Deep Learning

After an 'AI winter', a significant turning point came with the introduction of Alex Net, a deep convolutional neural network that dramatically outperformed traditional methods in the ImageNet competition of image classification. AlexNet demonstrated that, when combined with large-scale datasets and GPU acceleration, deep networks could achieve state-of-the-art performance in computer vision. This catalyzed a wave of interest in deep learning across various domains.

2015–2017: AlphaGo and AlphaGo Zero

AlphaGo marked the first time an AI defeated a human world champion in the ancient board game Go, a feat previously thought unattainable due to the game's complexity. This was followed by AlphaGo Zero, which was a major evolution in the sense that it learned without any human data, mastering the game purely through self-play.

These systems combined deep reinforcement learning with Monte Carlo Tree Search, illustrating the power of self-supervised learning and planning.

2020: AlphaFold

AlphaFold is a major achievement in science. It basically solved the decades-old protein folding problem by predicting 3D structures of proteins with near-experimental accuracy. AlphaFold was trained on a massive database of known protein structures, such as those found in the Protein Data Bank (PDB), enabling it to learn statistical patterns linking amino acid sequences to spatial conformations. This data-driven approach significantly outperformed traditional physics-based methods, which relied on the detailed modeling of biochemical interactions and molecular forces. By leveraging deep learning, AlphaFold achieved a level of predictive accuracy previously thought unattainable, transforming computational biology and offering new tools for scientific discovery in drug design, genomics, and disease understanding.

2022: ChatGPT and Large Language Models

ChatGPT, built upon the GPT-3 and GPT-4 transformer architectures, exemplifies the capabilities of large language models (LLM). Trained on vast text corpora and fine-tuned using reinforcement learning from human feedback, LLM can perform a range of natural language tasks with fluency and contextual awareness. A few months after their release, their widespread adoption sparked discussions about AI safety, bias, misinformation, and the broader implications of generative models.

These are some of the major steps which illustrate the extremely rapid progress of AI technology in the last 12 years.

3.3 Machine Learning and Neural Networks

Most of the remarkable progress in artificial intelligence over the last decade has been driven by the rise of deep neural networks (DNNs). These models consist of multiple layers of artificial neurons, where each layer transforms its input data through a combination of learned weights and nonlinear activation functions. This layered architecture allows the network to learn hierarchical representations of data, where each successive layer captures increasingly abstract features. An artificial neuron receives input values, multiplies them by corresponding weights, sums the results, and passes the sum through a non-linear activation function to produce its output. This process is repeated across layers, enabling the network to capture complex input–output relationships. Figure 3.1 illustrates a typical deep feedforward network (Fig. 3.1).

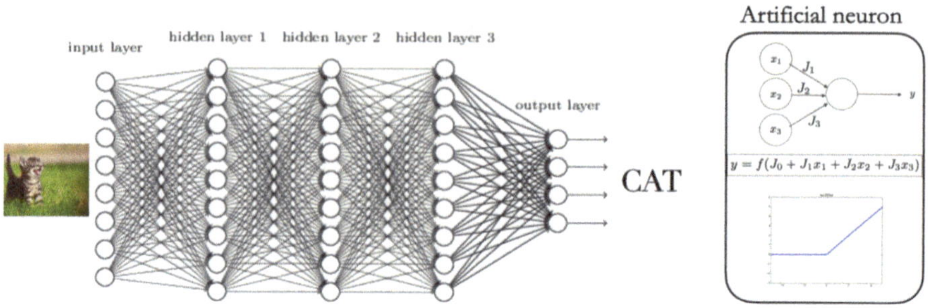

Fig. 3.1 Left: a deep feedforward neural network with multiple layers. Right: structure of an artificial neuron. *Source* deep_net.it; artificial_neuron.png

3.3.1 The Learning Task

A deep network is characterized by its general architecture (number of layers, width of the layers, types of nonlinearities in each layer), and by the set of weights which modulate the inputs to each of the neurons. Modern networks can have tens of billions of weights. Their values are obtained by supervised machine learning. This process consists in finding the weights such that the machine maps inputs (e.g., image pixels or word sequences) to outputs (e.g., categories or values), based on a labelled dataset. This process involves optimizing the weights to minimize a loss function that quantifies the difference between predicted and true labels. Because of the large number of weights to be fixed, the training needs to be done on massive datasets. The training process, while conceptually straightforward, is computationally intensive and typically carried out on high-performance GPUs or TPUs over days or even weeks.

3.3.2 Why Now? The Renaissance of Neural Networks

Feed forward neural networks in their present shape have existed since the 1980s, but several factors have recently contributed to their resurgence:

- Dramatic increase in computational power
- Availability of massive labelled datasets
- Advances in software frameworks and automatic differentiation libraries
- Accumulated empirical know-how for network design and training
- The emergence of generative models, capable of synthesizing data such as images and text.

The present technological breakthrough is largely due to the increase in computing power and availability of big data to train larger and larger networks. Mehonic and Kenyon illustrate that the amount of computing power used by latest AI software

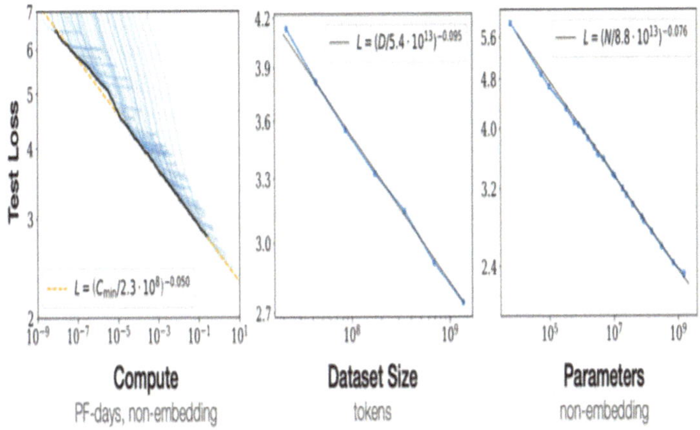

Fig. 3.2 Scaling laws for large language models, from Kaplan et al.: Language modeling performance improves as one increases the amount of computer, the dataset size, and the size of the network with power-law scaling involving small exponents. **Source**: Simple PowerLaws.pdf

has increased faster than exponentially in recent time. While it used to double every second year (Moore's law), since 2012 it seems to double every 3 to 4 months.

In addition, research such as Kaplan et al. has shown that performance improvements in large language models follow precise scaling laws, where predictive accuracy increases predictably with model size, dataset size, and compute. These scaling laws (see Fig. 3.2) guide model development and resource allocation. Because the scaling exponents are small, one needs to increase size, data and computer considerably in order to get good performance.

3.4 Generalization and the Black Box Problem

Well-trained deep networks present a paradox. On the one hand we know everything on the network: we can explicitly list and analyze all weights and activations, we know the value of each neuron when an input is presented to the network. But in practice the trained network is used as black box. You present a new input and you read in the output the classification "guessed", or "inferred" by the network. A big problem is that, in spite of knowing everything, we understand rather little of what the network actually does (Fig. 3.2).

The internal reasoning behind their decisions remains largely opaque. This opacity raises serious concerns:

- There is no guarantee that the model will perform well outside its training distribution.
- Interpretability is limited: although all learned parameters are known, the emergent behavior of the network is not well understood.

- The mechanism of generalization is not well understood. On the one hand, Zhang et al. demonstrated that deep networks can memorize completely random labels, yet still generalize well on real data, suggesting that traditional generalization theory is insufficient. Belkin et al. showed that overparameterized models can generalize effectively beyond the classical bias-variance tradeoff. These insights point to the complexity of generalization in high-dimensional, overparameterized models, and emphasize the need for new theoretical frameworks to better understand their behavior.

The lack of explainability in AI decision-making reduces trust and usability in critical applications such as healthcare, finance, and law enforcement. Unlike traditional algorithms that come with well-defined rules and guarantees, deep learning models operate without formal assurances of their performance. As AI becomes more deeply integrated into decision-making processes, the inability to explain outcomes, or develop some kind of logical or causal justification, poses significant risks. Without a theoretical foundation to explain these phenomena, AI remains an unpredictable tool: its prediction maybe statistically well founded, but we have no controls on possible outliers, which pose a serious problem in safety–critical domains.

To summarize, the disconnect between empirical success and theoretical understanding is one of the central challenges in modern machine learning. It is also one of the most important scientific challenges today, which connects deeply with fundamental questions in neuroscience and complex systems.

3.5 Societal and Security Implications

AI is increasingly influencing society in ways that raise major ethical and security concerns. I shall not address all the many issues related to the transformation of work, but I want to insist on three types of risks.

One of the primary risks is mass surveillance, where AI-driven monitoring systems threaten individual privacy and civil liberties. Governments and corporations leverage AI to track behaviors, analyze data, and even predict actions, often without adequate oversight or transparency. Already in 2019, Steven Feldstein was underlying that "*Around the world, AI systems are showing their potential for abetting repressive regimes and upending the relationship between citizen and state, thereby accelerating a global resurgence of authoritarianism*", underlying that "*it offers a number of particular benefits to authoritarian and illiberal regimes*". This tendency has become even worse in the last 5 years. It isn't a risk for the future, it is already well present. In several countries, police forces use face recognition device in order to screen and target demonstrators or political opponents. As mentioned by Gian Piero Siroli during this conference, there have been reports that AI has been used by Israeli Defense Force to target potential Hamas terrorists after October 7 attack, including locating their homes and destroying these homes and all inhabitants.

A second risk already at work is associated with one of the most remarkable developments of the last years in AI: generative models. In these models, instead of classifying, one attempts at generating fake data. Starting from a database, for instance a set of images of dogs, one aims at generating a new image of a dog which is not an image in the database (this would be memorization), but a new image of a new (non-existing) dog. A lot of progress has been made in recent time on generative models, and for instance generative diffusion techniques have reached a high level of performance on images, sound and video generation.

At the heart of these methods is the capacity of deep networks to interpolate in large dimensions: in some sense they can interpolate between the many images of dogs they have seen in the database to create a new image. In this case the risk of mischievous use of this technological breakthrough is obvious. AI can now be used as a mass factory of fake news. It is very hard to imagine how a democracy can function in such a context. This has been clearly summarized by Hanah Arendt: "*What makes it possible for a totalitarian or any other dictatorship to rule is that people are not informed; how can you have an opinion if you are not informed? If everybody always lies to you, the consequence is not that you believe the lies, but rather that nobody believes anything any longer*". AI-generated content, including deepfake videos and highly convincing fake news, can influence public opinion and elections, undermining trust in information. As AI becomes more capable, the potential for malicious use increases, necessitating stringent policies and frameworks to mitigate these risks. There are many occurrences of using such massive disinformation to destabilize other countries, or convince their public opinion of some fake situations. This could also very well be relevant in the context of potential use of nuclear weapons.

The third type of risk I want to stress is of primary concern, and deeply related to the topic of this conference. Deployment of AI in warfare presents profound ethical dilemmas. Lethal Autonomous Weapon Systems (LAWS) have the potential to conduct military operations without human intervention, raising questions about accountability and control. The ability of AI to make autonomous kill decisions challenges the traditional frameworks of international law and warfare ethics. While several military commands have announced ethical code decisions requiring a 'man in the decision loop' who can authorize the order to fire on a target, it will be very difficult to implement this position on the ground. The capacity of gathering and analyzing information very fast provides a decisive advantage on the battlefield. Clearly, the capacity of AI to analyze almost instantaneously a large amount of data from various sources and propose a tactical decision is used and will be used more and more. Implementing the ethical code under the pressure to make fast decisions will not be easy. This acceleration of the decision time is a global threat, particularly in the context of nuclear weapons.

3.6 Conclusions

AI is a powerful yet poorly understood tool. The rapid technological progress necessitates the development of a theoretical foundation to ensure safe and ethical deployment. Addressing interpretability, robustness, and societal impact remains crucial for future advancements. As AI continues to integrate into critical decision-making processes, interdisciplinary efforts in ethics, law, and technical research are needed to bridge the gap between capability and understanding. Only by fostering transparency and accountability can AI truly benefit society without unintended consequences. The challenge ahead lies not only in advancing AI technology but also in ensuring that it serves humanity in a manner that is responsible, ethical, and sustainable.

Furthermore, there is no way one can hope to stop AI proliferation: while training very large scale foundation LLM is costly, and so far limited to a few private companies, adapting AI tools to specific situations is relatively easy, and it is very hard to monitor or limit. Societies and regulations need to adapt to the accelerated evolution of AI, developing a culture and sets of tools which will be able to mitigate the risks.

Bibliography

1. H. Arendt, The New York Review of Books. **25**(15), 50–56 (1978). Interview by Roger Errera, originally conducted in 1974 (1978)
2. M. Belkin, D. Hsu et al., Reconciling modern machine-learning practice and the classical bias-variance trade-off. Proc. Nat. Acad. Sci. **116**(32), 15849–15854 (2019)
3. H.M. Berman, J. Westbrook et al., The protein data bank. Nucleic Acids Res. **28**(1), 235–242 (2000)
4. T.B. Brown, B. Mann et al., Language models are few-shot learners. Adv. Neural. Inf. Process. Syst. **33**, 1877–1901 (2020)
5. N. Chen, Y. Zhang et al., WaveGrad: estimating gradients for waveform generation (2019). arXiv:2009.00713
6. S. Feldstein, How artificial intelligence is reshaping repression. J. Democr. **30**, 40–52 (2019)
7. G.E. Hinton, T.J. Sejnowski, Optimal perceptual inference, in *Technical Report* (University of California, San Diego, Institute for Cognitive Science 1983)
8. J. Ho, A. Jain, P. Abel, Denoising diffusion probabilistic models, in *Neural Information Processing Systems. Conference on Neural Information Processing Systems* (2020)
9. J. Ho, T. Salimans et al., Video Diffusion models (2022). arXiv:2204.03458
10. J.J. Hopfield, Neural networks and physical systems with emergent collective computational abilities. Proc. Nat. Acad. Sci. **79**(8), 2554–2558 (1982)
11. J. Jumper, R. Evans et al., Highly accurate protein structure prediction with alphafold. Nature **596**(7873), 583–589 (2021)
12. J. Kaplan, S. McCandlish et al., Scaling laws for neural language models (2020). arXiv preprint arXiv:2001.08361
13. A. Krizhevsky, Sutskever et al., ImageNet classification with deep convolutional neural networks. Adv. Neural Inf. Proc. Syst. (NIPS) **25**, 1097–1105 (2012)
14. Y. LeCun, B. Boser, et al., Handwritten digit recognition with a back-propagation network. Adv. Neural Inf. Proc. Syst. (NIPS) 396–404 (1990)
15. A. Mehonic, A.J.Kenyon, Brain-inspired computing needs a master plan. Nature **604**(7904), 255–260 (2022)

16. M. Minsky, S. Papert, *Perceptrons: An Introduction to Computational Geometry* (MIT Press, 1969)
17. L. Ouyang, J. Wu et al., Training language models to follow instructions with human feedback (2022). arXiv preprint arXiv:2203.02155
18. F. Rosenblatt, The perceptron: a probabilistic model for information storage and organization in the brain. Psychol. Rev. **65**(6), 386–408 (1958)
19. D.E. Rumelhart, G.E. Hinton, R.J. Williams, Learning representations by back-propagating errors. Nature **323**(6088), 533–536 (1986)
20. D. Silver, A. Huang et al., Mastering the game of go with deep neural networks and tree search. Nature **529**(7587), 484–489 (2016)
21. D. Silver, J. Schrittwieser et al., Mastering the game of go without human knowledge. Nature **550**(7676), 354–359 (2017)
22. Y. Song, J. Sohl-Dickstein et al., Score-based generative modeling through stochastic differential equations, in *International Conference on Learning Representations* (ICLR, 2020)
23. C. Zhang, S. Bengio et al., Understanding deep learning requires rethinking generalization, in *International Conference on Learning Representations* (ICLR, 2017)

Chapter 4
Warfare at AI Speed and the Extended Meaningful Human Control Problem

Daniele Amoroso, Diego Mauri, and Guglielmo Tamburrini

Abstract As militaries increasingly incorporate Artificial Intelligence (AI) into their decision-making architecture, legal and ethical debates have primarily focused on Autonomous Weapon Systems (AWS). This paper argues that equally urgent challenges concern AI-enabled Decision-Support Systems (DSS-AI), even though these systems formally preserve human authority over the use of force. We develop the concept of extended meaningful human control (MHC) to address normative risks posed by DSS-AI in operational environments where time pressure, cognitive constraints, and system opacity converge to marginalize human agency in the decision-making loop. Our argument draws on two scenarios in which the integration of AI in military decision-making is linked to the scarcity of temporal resources—whether inherent (as in nuclear early warning systems) or induced (as in the alleged use of DSS-AI by the Israeli Defense Forces in Gaza). Both cases illustrate how taking critical military decisions at AI speed raises profound ethical and legal concerns. To resist the erosion of MHC, we propose three implementation pathways: human-centered system design, clearly distributed responsibility within command structures, and training programs to counter cognitive and institutional sources of automation bias. These measures should be codified in a multilateral treaty or, failing that, embedded in national doctrines and military manuals to resist the systemic drift toward "warfare at AI speed".

D. Amoroso
University of Cagliari, Cagliari, Italy
e-mail: daniele.amoroso@unica.it

D. Mauri
University of Palermo, Palermo, Italy
e-mail: diego.mauri@unipa.it

G. Tamburrini (✉)
University of Naples Federico II, Naples, Italy
e-mail: guglielmo.tamburrini@unina.it

P. Cotta-Ramusino et al. (eds.), *Nuclear Weapon Risks in a Problematic Time*, Springer Proceedings in Physics 358, https://doi.org/10.1007/978-3-032-29873-7_4

4.1 Introduction

Military powers are rapidly incorporating Artificial Intelligence (AI) technologies into their defense systems. These AI-enabled developments include Autonomous Weapons Systems (AWS) and systems supporting multiple facets of military decision-making, which range from strategic planning to operational support of battlefield actions.

Decision-support systems enabled by AI technologies (DSS-AI from now on) crucially differ from AWS in the control roles that are assigned to human operators. AWS can select and then proceed to engage targets without requiring any human intervention after their activation. By contrast, DSS-AI provide advice and other relevant information to human decision-makers. The latter are assigned by design the power to accept, revise, or reject the content of these inputs. In view of their autonomy in target selection and engagement, AWS have raised ethically and legally significant human control issues [4, 11, 13]. DSS-AI have been found to raise human control issues of a different kind [21], whose insurgence is not prevented by the fact that DSS-AI, unlike AWS, are subject by design to human judgment and gating functions.

In this contribution, we focus on military decision-making involving DSS-AI at the operational level, identifying major sources of human control hurdles in the narrow temporal resources that are often available in those decision-making contexts and in limitations of our capability to understand, predict precisely, and explain the behavior of AI systems in general, and DSS-AI in particular. This general claim is explored in connection with two different scenarios.

The first scenario (examined in Sect. 4.2) involves the use of DSS-AI in nuclear early warning. A major motivation for this proposal is that sensory processing at AI speed can help one achieve situational awareness more rapidly, thereby making more time available for downstream human decision-making in response to nuclear attack warnings. It is argued, however, that this expectation fails to properly consider the extra time that one must spend to verify the correctness of DSS-AI outputs. This extra time may offset and even reduce the time available for downstream human decision-making.

The second scenario (examined in Sect. 4.3) concerns the alleged use of DSS-AI made by the Israeli Defense Forces (IDF) to identify bombing targets in Gaza. Unlike the nuclear early warning scenario, which is characterized by an inherent scarcity of temporal resources, a problem of induced scarcity of temporal resources is denounced to arise in this case—more specifically, scarcity induced by hierarchical pressure to adapt the pace of human decision-making to the speed of AI information processing. It is argued (Sect. 4.3.1) that this induced scarcity may trigger human cognitive biases, which jeopardize the effectiveness of human control on DSS-AI advice.

In Sect. 4.4, it is claimed that in both conditions of inherent and induced scarcity of temporal resources, limitations of present AI technologies and human cognitive biases converge to erode the meaningful human control (MHC) that one should exert on military DSS-AI used in operational decision-making. And the erosion of MHC on DSS-AI is found to introduce new threats to the respect of international law, including International Humanitarian Law (IHL) and the law against war (*jus ad bellum* or *jus contra bellum*). While existing rules and principles of international law can already be construed as requesting MHC on DSS-AI, it is argued that States and other international actors must urgently countervail the ongoing erosion of MHC on DSS-AI, by sharing best practices based on human-centered control design and training, by introducing appropriate national and international regulation, and ultimately by checking the ongoing military race towards the objective of fighting wars at AI speed. Acting against the hunger for augmented speed in the battlespace is precisely what legal and ethical requirements currently impose. Section 4.5 concludes.

4.2 The Illusion of Buying Time: AI Speed and Human Control in Nuclear Early Warning

The major nuclear powers seem to converge on the position that decisions to use nuclear weapons must firmly remain in human hands. This is not only supported by self-evident political exigencies, but also inferable from the rules on the use of force in international relations and self-defense (*jus ad bellum*) and those of international humanitarian law (IHL, *jus in bello*), which severely restrict the threat or use of nuclear weapons. As the International Court of Justice famously held, indeed, the "use of nuclear weapons would generally be contrary to the rules of international law applicable in armed conflict", the only exception being their use "in an extreme circumstance of self-defence, in which the very survival of a State would be at stake" (ICJ, *Legality of the Threat or Use of Nuclear Weapons*, Advisory Opinion of 8 July 1996, Dispositif, para. 2D).[1]

However, nuclear states have not excluded several ways in which DSS-AI may contribute to nuclear command, control, and communication (NC3), including the identification of potential missile launches through sensor data processing, prelaunch activity detection, and the selection of actions to undertake in response to nuclear threats [22, 8]. In this section, we focus on the proposal to use DSS-AI for potential missile launch identification. The main rationale for this proposal to use DSS-AI in nuclear early warning is that AI processing is expected to reduce potential errors, and its speed is expected to buy more time for nuclear decision-makers within the narrow temporal windows that are available to them [19, 104, note 22]. These temporal windows are *inherently* narrow insofar as the time span typically elapsing from missile launch to impact is less than 30 min.

It is undoubtedly of paramount importance to make as much time as possible available to nuclear decision-makers within that inherently restricted temporal window; what is more, this should be 'quality' time, in which decision-makers may carefully

[1] It should be noted that the Court was deeply divided on this point, which was included in the Dispositif only thanks to the President's casting vote.

evaluate the real nature of the incoming attack and whether a nuclear response is due and in line with *jus ad bellum* requirements of immediacy, necessity and proportionality [20]. However, the relevant question is whether AI provides effective means to pursue this goal. Significant doubts about a positive answer to this question are fueled by reflecting on current limitations of AI technologies based on machine learning and on related human-AI interactions. Let us briefly recall here some of these limitations, which have played a central role already in the framework of ethical and legal debates about the human control of AWS [4]. These limitations notably include (i) the epistemic difficulty to go beyond statistical assessments of an AI system's accuracy and to predict its pointwise behaviors; (ii) AI errors that adversarial machine learning investigations have highlighted, and that the human cognitive system is unlikely to commit; (iii) the opacity of much AI information processing.

Consider item (i) under the assumption that a DSS-AI system was already tested and found to achieve highly accurate results in nuclear launch detection. These accuracy results do not exclude the occurrence of an infrequent mistake, such as a false positive detection of incoming nuclear missiles. Since the occurrence of such infrequent mistakes may trigger an unjustified use of nuclear weapons, the ensuing high risk for civilian populations and human civilizations demands systematic probing of the veracity of DSS-AI perceptual responses. But the time spent on this task may offset any temporal gain for human decision-makers that one may expect to flow from this use of DSS-AI in nuclear early warning.

Similar doubts are strengthened by considering item (ii), that is, the discovery of unexpected fragilities in AI information processing, leading to mistakes that AI systems may commit, and that the human perceptual system would unproblematically detect and avoid. These fragilities of AI information processing have been uncovered by means of adversarial machine learning techniques [7], thereby revealing perceptual situations in which human information processing is more robust than AI information processing. One cannot presently exclude that AI systems for nuclear early warning will make counterintuitive and potentially catastrophic errors of the same sort that adversarial machine learning highlighted in other critical application domains. Bringing these reflections to bear on international law, the issue emerges—which is admittedly little explored in doctrine—of the so-called "putative" self-defense, that is, the mistake of fact committed by a State that in good faith, but mistakenly, believes an armed attack is occurring against it, and thus reacts by resorting to force [15]. While the debate on whether genuine mistakes of fact may justify States' reactions concerns mostly cyberattacks, it seems to us that it affects also—and to a more troubling extent—DSS-AI in nuclear early warning, due to the catastrophic consequences that such mistakes of fact are likely to generate.

Let us finally consider item (iii). Arguably, assessing the veracity of DSS-AI responses may require human decision-makers to get some understanding of the reasons supporting those responses. But this task introduces additional temporal delays, amplified by the information processing opacity presently affecting DSS-AI and other AI systems based on machine learning and Deep Neural Networks (DNNs).

In conclusion, a reflection on limitations (i)–(iii) raises legitimate doubts on the prospects of leveraging AI processing speed to ease inherently scarce temporal resources in nuclear decision-making, insofar as this expectation overlooks that extra time must be spent to verify the correctness of DSS-AI outputs in this high-risk military domain. Such 'extra' time stands not only as an ethical requirement (making 'correct' choices in a highly critical domain) but also as a key tool to ensure respect for *jus ad bellum* requirements.

4.3 Inducing Time Pressure Through AI: IDF Uses of DSS-AI in Gaza

In this section, we turn to consider the military drive to leverage AI speed from a different perspective: from the proposal of buying more time for human decision-makers with the help of AI speed, we now examine proposals to leverage AI speed to increase the pace of operational decisions concerning actions in conventional battlefields.

A suitable scenario that enables one to explore these proposals is afforded by news reports concerning DSS-AI systems, known as *Habsora* (Gospel), *Where's Daddy* and *Lavender*, employed by the Israeli Defense Forces (IDF) in the armed conflict in Palestinian territories. Israel has resorted heavily to those systems to identify bombing targets in Gaza and its surroundings, in the aftermath of the massacre of innocent civilians perpetrated by Hamas within Israeli territory on October 7, 2023.

According to investigations carried out by the + 972/Local Call magazines and published between 2023 and 2024 [1, 2], the IDF employed these DSS-AI to generate lists of potential bombing targets in the Gaza Strip. A team of specialists in the IDF targeting division was appointed with the task of reviewing downstream the items included in AI-generated lists of potential targets, eventually approving or rejecting AI recommendations based on a legitimacy check. However, undisclosed sources in the IDF targeting division stated that the automation of the target generation task increased dramatically the targeting division's productivity in the way of potential target generation, moving "from 50 targets a year to 100 targets a day". While this stunning increase does not per se run in contravention with existing international rules, it must be assessed whether this drive to augmented speed in the battlespace risks fueling behaviors that are not in line with legal and ethical requirements.

The increased machine productivity in target generation gave rise to a decision-making bottleneck. According to an anonymous source, human operators perceived psychological pressure to align their filtering decisions to the pace of machine targeting suggestions. Indeed, this source complains about the assembly line character imposed on the filtering task ("it really is like a factory"), about the lack of sufficiently extended time frames to overview machine suggestions and to provide considered judgments about legal compliance ("we work quickly and there is no time to delve deep into the target"). Related to this, the same anonymous source mentions the introduction, higher up in the military hierarchy, of new criteria to evaluate the performances of targeting division members, which prize increased output volumes of approved targeting suggestions ("the view is that we are judged according to how many targets we manage to generate") [1]. Consistently with this view, target selection procedures were drastically simplified: "One source stated that human personnel often served only as a 'rubber stamp' for the machine's decisions, adding that, normally, they would personally devote only about '20s' to each target before authorizing a bombing—just to make sure the Lavender-marked target is male." [2].

In the scenario described in this investigative news report, military commanders expect human control to be exerted under more demanding temporal constraints, intimate their subordinates to adapt the pace of human decision-making to increased machine productivity, and provide incentives when the acceptance rate on DSS-AI suggestions goes up.

Let us now examine this scenario from the perspective of MHC and the related normative requirement of protecting MHC functions from situations in which human control boils down to little more than clerical rubber-stamping of machine suggestions.

The described circumstances suffice to contend that the practice of DSS-AI in the Israel-Palestine conflict fails to comply with the principle of precautions in attacks, one of the cornerstones of the law of armed conflict. What is more, it is known that those systems make what are regarded as "errors" in approximately 10% of cases, occasionally marking individuals who have merely a loose connection to militant groups, or no connection at all [2]. It follows that Israel accepts not only the hazard, but even the virtual certainty of engaging impermissible targets—a conduct that breaches existing rules on distinction and proportionality in the conduct of hostilities, giving rise to both international responsibility of the State and, to the extent that such conduct qualifies as war crimes such as attacks against the civilian population and excessive collateral damage under Art. 8(2)(b)(i) and (iv) of the ICC Statute, international criminal responsibility for individuals involved [5, 14].

4.3.1 Automation Bias and Related Human Decision-Making Heuristics

It was noted in Sect. 4.2 that in situations brought out by adversarial machine learning, human and AI perceptual processing show some complementary strengths and weaknesses. In some circumstances, the human perceptual system appears to be more robust than AI perceptual processing. This observation supports the claim that MHC must play the prospective responsibility role of a fail-safe agent, contributing to some extent to counteracting possible effects of failure anticipated by adversarial machine learning. However, the performance of this fail-safe function is likely to be jeopardized by the automation bias onset. The expression "automation bias" designates a propensity of human decision-makers to defer to recommendations advanced by DSS systems despite warning signals or retrievable information to the contrary.

In the above IDF scenario, the onset of the automation bias may be facilitated by the so-called "authority bias", that is, the tendency to attribute greater value to the opinion of an authority figure and to be more influenced by that opinion [16]. Here, the authority bias may invite subordinates to emulate behaviors of lenient attitudes towards DSS advice adopted by role models in the targeting division and incentivized by hierarchical superiors. The reluctance to question "algorithmic authority" may additionally stem from a different, albeit related phenomenon: a DSS-AI operator would indeed be strongly discouraged from overriding the system's recommendation by the fact that he or she "will be personally blamed if the override goes wrong" with likely negative career consequences, while the failure to reject a harmful recommendation leaves open the possibility "to blame the AI for any difficult decision with a negative outcome" [8, 317].

Moreover, automation bias was experimentally probed in aviation and other time-critical control domains [10]. Anecdotal evidence of automation bias in time-critical military situations of DSS use has been widely reported, including friendly fire episodes in the first Iraqi war induced by wrong DSS detection of approaching enemy fighter jets. A psychological mechanism that has been hypothesized to pave the way to automation bias in time-critical domains has to do with the dual nature of human decision-making.

Dual models of human decision-making identify a significant source of human judgment perturbation in the psychological pressure to meet tight deadlines under severe cognitive load [25]. In this regard, Daniel Kahneman introduced an admittedly simplifying distinction between two distinct cognitive "systems", respectively subsuming heuristic and more analytic decision-making [12]. System 1 is a blanket term collecting heuristic decision-making processes that tend to be faster, mostly automatic, default and often emotionally laden. System 2 embraces more reflective and analytical decision-making processes that are usually slower and less readily activated. The two systems operate concurrently, but not always cooperatively. Options that are more readily selected by System 1 are often accepted by default, and concurrent efforts by System 2 are disregarded, especially when there is limited time to decide what to do.

To illustrate, consider the so-called WYSIATI (What-You-See-Is-All-There-Is) heuristics, identified and experimentally probed by Daniel Kahneman and Amos Tversky. WYSIATI is a System 1 heuristics reflecting, according to Kahneman, a "remarkable asymmetry between the ways our mind treats information that is currently available and information we do not have". People tend to focus on what is clearly visible to them, neglecting information that is not known or not presently retrieved from memory: "Information that is not retrieved (even unconsciously) from memory—Kahneman observed—might as well not exist." One jumps to conclusions based on this asymmetry and goes on to construe "the best possible story that incorporates ideas currently activated", endorsing by default the (often but not invariably correct) belief that only this information is relevant [12, 85]. The WYSIATI heuristics induces people to focus on what is clearly visible to them and to treat as non-existent information that is not known or not presently retrieved from memory. WYSIATI may contribute to predict or explain tendencies to conform to DSS-AI advice. In particular, the lists of targets generated by the DSS-AI allegedly used by the IDF in Gaza are readily visible to human controllers, whereas considerable cognitive effort might be needed on their part to carefully assess (and, in case, reject) them on rational grounds and to propose different courses of action. The anchoring heuristic strategy attributed to System 1 may further contribute to explain the tendency to conform to DSS-AI suggestions, insofar as an individual's decisions turn out to be influenced by an initial reference point or anchor. In this case, the anchor is provided by DSS-AI advice and the human controller's subsequent evaluations and conclusions tend to stay closer to the anchor than they would have been otherwise.

4.4 The Why and How of Extended MHC on AI-Enabled Warfare

A major source of what we call here the extended MHC problem is AI speed and its influence on human decision-making supported by DSS-AI. It has been suggested that both *inherent* and *induced* scarcity of temporal resources in operational decision-making involving DSS-AI significantly hinder the ability of human operators to intervene in the process.

Military deployment of these systems, in other terms, tends towards "a dramatically reduced decision-making cycle as the capabilities in this area improve" [18]. In this context, the operational tempo of decision-making is no longer dictated by human limits. It is rather adapted to the accelerated outputs of machines, with the human role increasingly being reactive rather than deliberative. Commanders who once had the luxury of measured deliberation will now be "forced to [make decisions] in seconds," particularly when facing adversaries leveraging AI to accelerate battlefield tempos [9].

The fear of lagging behind in the arms race toward hyper-accelerated decision-making is already influencing strategic doctrines among the world's leading military powers [23]. The competitive pressure to match or surpass the adversary's DSS-AI capabilities, as well as the ever-lasting doctrine of nuclear deterrence (now entering an unprecedented phase due to its combination with AI systems), lead to a global landscape situation where deliberation time itself becomes weaponized.

These developments point to a growing and pervasive normative challenge in contemporary military practice: the need to ensure meaningful human control in the use of DSS-AI. Unlike traditional discussions focused on autonomous weapon systems, this problem arises even when humans remain formally in the loop but operate under conditions that may undermine deliberation and oversight. Such extended MHC problem can be distinguished into two parts. The first concerns the *why*, that is, the extant motivations for requiring MHC on DSS-AI. The second part of the problem has to do with the *how* of such MHC, that is, with the ways military uses of DSS-AI must be harnessed to ensure MHC.

To begin with the *why*, one can draw on previous analyses of the ethical and legal motivations for requiring MHC in the use of Autonomous Weapon Systems (AWS). As suggested elsewhere, human agents are normatively required to play three key roles in weapon system operations: as fail-safe actors, accountability attractors, and moral agency enactors [4, 24].

Among these roles, the fail-safe function stands out as particularly crucial in the case of DSS-AI. It posits that human oversight is necessary to detect and correct failures that would otherwise go unaddressed by the system itself, especially when these failures stem from adversarial misclassifications, erroneous correlations, or incomprehensible black-box outputs.

One might argue that this role is redundant in the DSS-AI context, since a human is always present to take the final decision. Yet, as discussed in Sect. 4.3.1, the practical conditions of decision-making—marked by extreme time pressure, information overload, and hierarchical incentives—often preclude meaningful verification or dissent. What eventually results from the present analysis is a procedural façade of control, in which the presence of a human decision-maker cannot fulfill the underlying moral and legal function of serving as a check on system failure.

In this sense, the core concern articulated by Article36 with regard to AWS in its ground-breaking analyses on MHC—namely, the distinction between meaningful and perfunctory human control—resurfaces with full force in the DSS-AI domain [6]. If the human role does not include the capacity and conditions to act as a fail-safe, then the ethical and legal demands associated with human control remain unmet, even if a human technically signs off on each action.

Having outlined the *why* of MHC over DSS-AI, we now turn to the more challenging question of *how* such control might be practically implemented in AI-enabled military contexts. A valuable starting point in this regard is offered by the ICRC's 2021 position paper on autonomous weapon systems [11]. While the document does not explicitly use MHC terminology, it provides one of the more authoritative attempts to translate normative concerns for human control into concrete operational criteria. As such, it offers not only a regulatory benchmark for AWS, but also a conceptual scaffold for designing control mechanisms in the case of DSS-AI.

In particular, the ICRC articulates the following requirement for ensuring appropriate human oversight:

> (MHC_1) Human decision-makers must be in a position to assess whether AWS operation is constrained – by design or situation of use – to objects that are legitimate military targets, and to intervene as required. This presupposes limits on duration, scope, and scale of use, and requires provisions for proper human supervision, judgement, timely intervention, or deactivation.

Although this condition is formulated in relation to systems that autonomously apply force, its underlying structure is directly relevant to the challenges posed by DSS-AI. Here, the primary risk is not kinetic action, but epistemic distortion—namely, the influence of machine-generated inferences on human situational awareness and decision-making. Extending the ICRC's approach to the DSS-AI domain, one may envisage a parallel requirement:

(MHC_2) Human decision-makers must be in a position to assess the veracity of DSS-AI perceptual inferences, to form pondered and well-informed judgments about their recommendations, to evaluate the likely real-world effects of actions undertaken on the basis of such inferences and recommendations, and to filter DSS-AI information and advice.

Translating the condition expressed in (MHC_2) into concrete operational measures requires addressing the interplay of cognitive, institutional, and procedural factors that determine whether human control is merely formal or functionally effective. The following elements are proposed as essential to the implementation of extended MHC in the DSS-AI domain:

- *Adopting a human-centered approach grounded in cognitive ergonomics.* The design of DSS-AI interfaces and workflows must take into account human cognitive limitations, especially under conditions of stress and time pressure. Insights from cognitive ergonomics highlight the importance of mitigating cognitive biases that are exacerbated by the accelerated tempo of AI-supported decision-making. In particular, systems should be designed to slow down decision points where high-stakes judgments are required, or to flag situations where overreliance on machine inferences is likely. This includes ensuring that output is presented in interpretable formats and that users are not cognitively overloaded by irrelevant or excessive information [17].
- *Ensuring a clear distribution of responsibilities along the chain of command and control.* The allocation of roles and responsibilities in the use of DSS-AI must be explicitly defined, including at the design, deployment, and operational stages. This requirement is rooted in both ethical imperatives and established legal frameworks, particularly those of international humanitarian law (IHL) and international criminal law (ICL), let alone *jus ad bellum* rules that are of paramount importance in scenarios involving AI-DSS nuclear warning systems. Discussions around human responsibilities for attacks against civilians allegedly carried out by the IDF on the advice of DSS-AI illustrate the legal and moral necessity of clarifying who is responsible for what in AI-assisted decision-making processes.
- *Promoting suitable training programs.* One key challenge to MHC is the formation of prior beliefs about AI capabilities, which are more and more often shaped by technological hype and related marketing narratives. The prevailing discourse around DSS-AI tends to emphasize superhuman accuracy and efficiency, while obscuring system limitations such as brittleness and lack of contextual understanding. Moreover, pre-conceptions formed through experience may also result in selective adherence, whereby users accept DSS recommendations only when they confirm preexisting intuitions or stereotypes [3]. Training should aim to correct these fallacies and prepare human operators to critically engage with DSS outputs. In other terms, it must focus on fostering reflective skepticism, not just "procedural familiarity" with the system.

Together, these three elements—human-centered design, clearly distributed responsibilities, and reflective training—constitute the operational core of extended MHC in DSS-AI environments. Their effectiveness will depend on their incorporation into formal regulations, standard operating procedures, and international best practices. Codifying these measures at both national and international levels is essential to resisting the accelerationist competition that currently governs much of AI adoption efforts in military contexts. Without such regulatory embedding, the very speed that gives DSS-AI its tactical appeal will continue to erode the conditions that are necessary to preserve and enhance human control.

4.5 Conclusions

This paper has explored the emerging challenge of extending the scope of the MHC requirement to cover AI-enabled decision-support systems in military operations. It has been argued that the problem of MHC is not confined to the automation of lethal force, but applies also to decision-making processes increasingly shaped by DSS-AI. We thus proposed an extended MHC model that identifies human control not simply with presence in the decision loop, but with the capacity to exercise judgment under conditions that are cognitively and institutionally conducive to meaningfully perform that task.

We argued that DSS-AI—despite being functionally different from AWS—raise similar normative concerns. The issue is not whether force is applied autonomously or manually, but whether critical functions of target identification and engagement remain under conditions that allow for genuine human deliberation. As Article36 foresightedly put it more than a decade ago, what matters is not who *presses the button*, but whether the decision-making context permits meaningful human agency in interpreting the information that leads to that action.

To address the *how* of extended MHC in AI-supported military operations, we outlined a set of operational recommendations. These include adopting a human-centered design approach informed by cognitive ergonomics; ensuring clearly distributed responsibility throughout the chain of command; and implementing training programs that promote critical engagement with AI systems.

Looking forward, a number of steps might profitably be taken to bring the extended MHC problem into the fold of global governance.

First, the issue should be integrated into key arms control and disarmament forums, including the Non-Proliferation Treaty (NPT) review process and high-level initiatives such as the Summit on Responsible AI in the Military Domain (REAIM) [22].

Second, a comprehensive and empirically grounded inquiry is needed to identify where, how, and under what conditions AI systems are already being deployed in military contexts that challenge the feasibility of extended MHC. This diagnostic effort should inform regulatory priorities and guide the development of accountability structures.

Third and finally, international legal instruments should be considered to regulate military applications of AI beyond (and *before*) AWS, using extended MHC as a foundational principle. Such an initiative might take the form of a legally binding treaty dedicated to arms control in the AI domain.

One must be realistic, though, about the geopolitical headwinds facing multilateral disarmament efforts today, ranging from the deterioration of relations between military powers (most notably the US and China), to renewed military investments around the world (including in the EU under the banner of "Readiness2030") and the worrisome withdrawal of some States from cornerstone treaties such as the Anti-Personnel Mine Ban Convention.

A second-best pathway lies in embedding extended MHC principles within national doctrines, especially IHL manuals and operational protocols. This could lay the groundwork for the eventual emergence of customary international norms or help coalesce, at least in the medium term, the consensus required for future treaty negotiation.

References

1. Y. Abraham, A mass assassination factory: inside Israel's calculated bombing of Gaza, in *+972 Magazine* (2023). https://www.972mag.com/mass-assassination-factory-israel-calculated-bombing-gaza
2. Y. Abraham, Lavender: the AI machine directing Israel's bombing spree in Gaza, in *+972 Magazine* (2024). https://www.972mag.com/lavender-ai-israeli-army-gaza/
3. S. Alon-Barkat, M. Busuioc, Human–AI interactions in public sector decision making: "automation bias" and "selective adherence" to algorithmic advice. J. Public Adm. Res. Theory **33**, 153–169 (2023)
4. D. Amoroso, G. Tamburrini, Toward a normative model of meaningful human control over weapons systems. Ethics Int. Aff. **35**(2), 245–272 (2021)
5. D. Amoroso, Sistemi di supporto alle decisioni basati sull'IA e crimini di guerra: alcune riflessioni alla luce di una recente inchiesta giornalistica. Diritti umani diritto internazionale **18**(2), 347–368 (2024)
6. Article36, Structuring debate on autonomous weapons systems, in Briefing Paper, Geneva 14–15 November (2013)
7. B. Biggio, F. Roli, Wild patterns: ten years after the rise of adversarial machine learning. Pattern Recogn. **84**, 317–331 (2018)
8. N. Bostrom, E. Yudkowsky, The ethics of artificial intelligence, in *The Cambridge Handbook of Artificial Intelligence*, eds. by K. Frankish, W.M. Ramsey (Cambridge University Press, Cambridge, 2014), pp. 316–334
9. J. Brendarn, A. Kulkarni, Fighting for seconds: warfare at the speed of artificial intelligence. Modern War Inst. West point (2023). https://mwi.westpoint.edu/fighting-for-seconds-warfare-at-the-speed-of-artificial-intelligence/
10. M.L. Cummings, Automation bias in intelligent time critical decision support systems, in *Decision Making in Aviation*, ed. by D. Harris (Routledge, London, 2015)
11. International Committee of the Red Cross, Position paper on Autonomous Weapons Systems. Geneva (2021)
12. D. Kahneman, *Thinking, Fast and Slow* (Farrar, Straus and Giroux, New York, 2012)
13. D. Mauri, *Autonomous Weapons Systems and the Protection of the Human Person. An International Law Analysis* (Edward Elgar, Cheltenham/Northampton, 2022)
14. D. Mauri, Numeri, persone, umanità. Sistemi di supporto alle decisioni umane in campo militare da parte dell'IDF e diritto internazionale umanitario. Diritti umani diritto internazionale **18**(2), 329–346 (2024)
15. M. Milanović, Mistakes of fact when using lethal force in international law: part II. EJIL:Talk! (2020). https://www.ejiltalk.org/mistakes-of-fact-when-using-lethal-force-in-international-law-part-ii/

16. S. Milgram, Behavioral study of obedience. J. Abnorm. Soc. Psychol. **67**(4), 371–378 (1963)
17. M. Draper et al., Human Systems Integration for Meaningful Human Control over AI-based systems (NATO, STO Technical Report, 2025). https://www.sto.nato.int/document/human-systems-integration-for-meaningful-human-control-over-ai-based-systems/
18. G. Moy et al., *Technical Report. Recent Advances in Artificial Intelligence and Their Impact on Defence* (Defence Science and Technology Group, Canberra, 2020). www.dst.defence.gov.au/sites/default/files/publications/documents/DST-Group-TR-3716_0.pdf
19. NSCAI (National Security Commission on Artificial Intelligence), Final Report (2021). https://assets.foleon.com/eu-central-1/de-uploads-7e3kk3/48187/nscai_full_report_digital.04d6b124173c.pdf
20. C. O'Meara, *Necessity and Proportionality and the Right of Self-Defence in International Law* (Oxford University Press, New York, 2021)
21. E.S. Probasco et al., AI for military decision-making. Harnessing the advantages and avoiding the risks, in *Issue Brief* (Center for Security and Emerging Technology, 2025). https://cset.georgetown.edu/publication/ai-for-military-decision-making/
22. A. Saltini, Assessing the implications of integrating AI in nuclear decision-making systems, in *Policy Brief* (European Leadership Network, 2025). https://europeanleadershipnetwork.org/policy-brief/assessing-the-implications-of-integrating-ai-in-nuclear-decision-making-systems/
23. F. Sauer, The military rationale for AI, in *Armament, Arms Control and Artificial Intelligence. The Janus-faced Nature of Machine learning in the Military Realm*, eds. by T. Reinhold, N. Schörnig (Springer, Cham, 2022), pp. 27–38
24. P. Scharre, Centaur warfighting: the false choice of humans vs. automation. Temple Int. Comp. Law J. **30**(1), 151–165 (2015)
25. F. Strack, R. Deutsch, The duality of everyday life: dual-process and dual system models in social psychology, in *APA Handbook of Personality and Social Psychology, Vol. 1. Attitudes and Social Cognition*, eds. by M. Mikulincer et al. (APA Books Washington DC, 2015), pp. 891–927

Chapter 5
Risks of Nuclear War: A Brief Partial Overview

Francesco Lenci

Abstract I will try to give a brief partial overview of the risks of nuclear war at a time of severe changes in the international scenario and extremely dangerous uncertainty for international security.

I wish to start by quoting a relatively recent statement from Pugwash officials: "*The present conflicts see the participation, in various roles, of countries that possess nuclear weapons: Russia, Israel, USA, UK, France* (and recently North Korea). *The most serious nuclear risk today is related to the war in Ukraine. In particular, Russian authorities and commentators have spoken several times about the possibilities of their use of nuclear weapons, if a direct attack to their territory is performed. A reference to the possible use of the so-called tactical nuclear weapons is often made. Today the probability of an actual use of nuclear weapons may still be a limited one. Nevertheless, even a very small probability of the use of nuclear weapons should be seriously considered as an unacceptable risk.*"

I am personally convinced that the use of even a few tactical nuclear weapons could rapidly escalate to a larger nuclear conflict. The dramatic consequences of a nuclear war have been discussed many times by scientists. As the Russel-Einstein Manifesto of 1955 reminded us: "*Shall we choose death, because we cannot forget our quarrels? We appeal, as human beings, to human beings: Remember your humanity, and forget the rest.*"[1]

In the 1980s, researchers showed that a nuclear war could cause vast amounts of smoke to spread around the globe, blocking out sunlight and turning summers into winters. Today's more sophisticated climate models show that the earlier research underestimated this impact.

There are about 12,000 nuclear weapons in the world, unevenly distributed among nine states (about 5000 USA and about 5500 Russia), some of them hundreds of times

[1] https://pugwash.org/1955/07/09/statement-manifesto.

F. Lenci (✉)
SICA International Security and Control Arms Committee, Accademia dei Lincei, Rome, Italy
e-mail: francesco.lenci@ibf.cnr.it

P. Cotta-Ramusino et al. (eds.), *Nuclear Weapon Risks in a Problematic Time*, Springer Proceedings in Physics 358, https://doi.org/10.1007/978-3-032-29873-7_5

more powerful than those which destroyed Hiroshima and Nagasaki. The use of just a few hundred could decimate the Earth's population, as shown by the three International Conferences on the Humanitarian Impact of Nuclear Weapons organized in Norway, Mexico, and Austria (2013 and 2014).[2]

With geopolitical tensions escalating, the risk of nuclear warfare is to its highest point in decades. As António Gutierrez, Secretary-General of the United Nations, said: "*There is one path—and one path only—that will vanquish this senseless and suicidal shadow, once and for all. We need disarmament now*".

In February 2024, China urged states with the largest nuclear arsenals to negotiate a treaty on no-first-use of nuclear weapons against each other. A roadmap or timetable should and could also be defined for an international legal instrument that would protect non-nuclear-weapon states from the threat of nuclear weapons.

I believe we should applaud to the awarding of the 2024 Nobel Peace Prize to Nihon Hidankyo, the organization of the *hibakusha* (the survivors of Hiroshima and Nagasaki) founded in 1956, "*for its efforts to achieve a world free of nuclear weapons and for demonstrating through witness testimony that nuclear weapons must never be used again.*"[3]

The risks associated with nuclear weapons include nuclear terrorism, miscalculations, false alarms, and escalation. A nuclear war has nearly occurred several times by accident, as described, for example, by Eric Schlosser in his 2016 *New Yorker* article on the "*Third World War by Mistake.*"[4]

Emerging technologies, such as hypersonic missiles, artificial intelligence and developments in cyber and outer space domains have created new risks. Efforts to incorporate artificial intelligence systems into nuclear weapons control raise the possibility of catastrophic miscalculations by accelerating response times and removing the human intuition and hesitation that have, on several occasions, helped prevent World War III. Consider the example of 1983: Colonel Petrov was the duty officer at the command center for the Oko nuclear early-warning system, when the system reported that a nuclear attack had been launched from the United States. Petrov judged the report to be a false alarm (as it indeed was) and prevented a retaliatory nuclear attack, thereby averting a large-scale nuclear war. This fact was revealed in 1998 with the publication of the memoirs of General Votintsev.

Interestingly, I appears that Ukraine, due to its full-scale war against Russia, does not plan to regulate AI development in the defense sector during the conflict (Minister Fedorov, June 2024).

All nuclear-armed states are spending enormous sums to expand and modernize their arsenals. According to SIPRI 2024 Yearbook, total global military expenditure reached about $2400 billion in 2023, an increase of 6.8% in real terms from 2022. Of the roughly 12,000 warheads, around 2100 are kept in a state of high operational alert on ballistic missiles—meaning just a few minutes to launch and less than about half

---

[2] https://www.reachingcriticalwill.org/disarmament-fora/hinw; https://www.icrc.org/en/statement/un-idenw.

[3] On the award of the 2024 Nobel Peace Prize, Stockholm, 11 October 2024.

[4] https://www.newyorker.com/news/news-desk/world-war-three-by-mistake.

an hour to reach their targets. According to Russian politician Volodin, the Russian Sarmat missile could reach Strasbourg in about 3 min). Nearly all of these 2100 warheads belong to Russia and the USA, but for the first time it seems that China has placed some warheads on high operational alert.

In 2024, the Bulletin of the Atomic Scientists once again set the Doomsday Clock at 90s to midnight, as humanity continues to face an unprecedented level of danger. Now, in 2025, the clock is at 89s from midnight.[5]

Leaders and citizens around the world should treat this as a stark warning and respond with urgency.

At the time of the Conference, in late October 2024, an end to Russia's war against Ukraine still seemed distant. And even though the conflict had been—and remained—a conventional one, it had also become very much a nuclear crisis. At the end of September 2024, President Putin announced possible changes to Russia's nuclear doctrine, stating: "*It is proposed that aggression against Russia by any non-nuclear state, but with the participation or support of a nuclear state, be considered as their joint attack on the Russian Federation.*" He added, "*The country's nuclear arms are the most important guarantee of security of our state and its citizens.*"

North Korea continued building nuclear weapons and long-range missiles. Nuclear expansion in Pakistan and India continued without pause or restraint. The hostility between the two Koreas, and the tense relations between China and Taiwan, remained critical.

The tragic situation in Gaza, Lebanon, Israel, Iran, and Yemen had the potential to escalate into a wider Middle Eastern conflict, posing unpredictable threats both regionally and globally. Let me quote Toshiyuki Mimaki, co-chair of Nihon Hidankyo and this year's Nobel Peace Prize laureate, said—comparing the situation in Gaza and Israel to that of Japan in 1945—"*Especially in places like Israel and Gaza, children are being covered in blood and living every day without food, having their schools destroyed, stations destroyed and bridges destroyed*" he said, appearing to fight back tears.

In August 1985, the *Bulletin of Atomic Scientists* published a paper by Józef Rotblat the only scientist to leave the Manhattan Project at the end of 1944, once it became clear that Nazi Germany would not be able to build nuclear weapons. Rotblat, who later co-founded the Pugwash Conferences (awarded the Nobel Peace Prize in 1995), concluded his paper with these words: "*After 40 years one question keeps nagging me: have we learned enough not to repeat the mistakes we made then? I am not sure even about myself. Not being an absolute pacifist, I cannot guarantee that I would not behave in the same way, should a similar situation arise. Our concepts of morality seem to get thrown overboard once military action starts. It is, therefore, most important not to allow such a situation to develop. Our prime effort must concentrate on the prevention of nuclear war, because in such a war not only morality but the whole fabric of civilization disappears. Eventually, however, we must aim at eliminating all kinds of war.*"

---

[5] https://thebulletin.org/doomsday-clock/2025-statement/.

Chapter 6
The Approaches to Reducing Nuclear Risks

Wuwen Zhao

Abstract This paper classifies the risk of nuclear use into two categories: intentional and unintentional. It emphasizes the importance of preventing risk from the perspective of its source and puts forward three suggestions for effectively preventing nuclear use. First, mutual respect for core interests. Second, reducing the role of nuclear weapons in national security policy. Third, extending the time for decision-making on the use of nuclear weapons.

6.1 Introduction

Today's international system is characterized by intensified competition among major powers and geopolitical conflicts. The future is full of uncertainty. Cooperation on nuclear arms control is becoming increasingly difficult, and the risk of nuclear weapons being used is growing. Almost sixty-three years ago, the Cuban Missile Crisis erupted, almost pushing the United States and the Soviet Union into nuclear war. Since then, the world has become acutely aware of the enormous risk of conflict escalating into nuclear war. Today, we do not need another Cuban missile crisis to highlight the importance of nuclear risk. As the theme of the Amaldi conference puts it, we are in a "problematic time", and answering the question of how to effectively prevent the risk of nuclear weapons being used is important if we want to achieve Mr. Amaldi's goals. [1].

6.2 Main Types of the Risk of Using Nuclear Weapons

The risks of using nuclear weapons can be categorized into two main types: unintentional and intentional.

There are three main types of risks of unintentional nuclear weapons use:

W. Zhao (✉)
Center for Strategic Studies, China Academy of Engineering Physics, Beijing, China
e-mail: zhww@caep.cn

P. Cotta-Ramusino et al. (eds.), *Nuclear Weapon Risks in a Problematic Time*, Springer Proceedings in Physics 358, https://doi.org/10.1007/978-3-032-29873-7_6

The first is the risk of unauthorized use of nuclear weapons.
The second is the risk of accidental use due to human error or technical failure.
The third is the risk of unintentional nuclear escalation due to misunderstandings and miscalculations. This risk is becoming increasingly obvious and serious against the backdrop of fierce competition between major powers and the complexity of regional security issues.

In terms of the risk of intentional nuclear use, the most realistic form today is the deterrence of nuclear and non-nuclear attacks through the first use of nuclear weapons.

6.3 Measures of Preventing the Risk of Using Nuclear Weapons

Whether the risk of nuclear use is intentional or not, it could eventually lead to the use of nuclear weapons, which would be an unbearable risk and a great catastrophe for all humankind.

Taking effective measures to avoid creating risks is of the utmost importance. Preventing crises is more important than managing them. Preventing the risk of nuclear weapons being used should address the root of the problem, and practical and effective measures should be taken to prevent nuclear risks. For example, it is irresponsible for some great powers to carry out provocative actions while emphasizing the need for crisis management.

To avoid the risk of nuclear weapons being used, it is necessary to make progress in three key areas.

(a) **Mutual Respect for Core Interests**

The major powers should respect each other, strengthen strategic dialogue and communication, and enhance understanding through mutual trust and benefit.

Great Powers should respect each other's core interests, such as sovereignty and territorial integrity, and avoid testing or threatening each other's "red lines" or "bottom lines" by force.

Avoiding each other's "red lines" and "bottom lines" is the most important factor in preventing conflicts and the risk of escalation to the use of nuclear weapons. If one party ignores the other's clear signals and continues to challenge their 'red line' or 'bottom line', it is difficult to describe the ignoring party as 'non-intentional', and the nuclear risk due to confrontation would increase significantly. The nuclear risk due to confrontation would then be greatly increased.

(b) **Effective Reduction of the Role of Nuclear Weapons**

For more than 60 years, China has consistently pursued a No-First-Use policy. This policy has been invaluable in reducing the risk of nuclear weapons use. It is an important symbol of the diminishing role of nuclear weapons in national security policies. It is also in line with the spirit of the Joint Statement of the five nuclear weapon states in 2022, which asserts that "nuclear war cannot be won and must never be fought". Adhering to the No-First-Use policy will allow major nuclear states to reduce their reliance on nuclear weapons and significantly decrease the risk of their deployment.

During his election campaign, President Biden made it clear that he wanted to reduce the role of nuclear weapons in national security policy and realize the "sole purpose" of nuclear weapons [2]. I personally believe that adopting a "sole purpose" approach would facilitate a smooth transition to No First Use of nuclear weapons, greatly reducing the risk of their deployment. Unfortunately, President Biden has not maintained his campaign views since taking office.

(c) **Extension of the Decision-making Time for the Use of Nuclear Weapons**

Even during the Cold War, Robert McNamara—then United States Secretary of Defense—advised two United States presidents that if information was received that a nuclear weapon had been detonated on US territory, the president should refrain from taking any retaliatory action and wait 24 hours for confirmation. This advice made it clear that all measures should be taken to avoid hasty decisions [3]. McNamara recognized that no adversary could disarm the United States, yet he was still aware of the dangers of hasty decision-making and preferred a "wait-and-see" policy [4].

In the face of the risks posed by false warnings, extended decision-making time is conducive to effective risk prevention. It can prevent misjudgments caused by technical errors and false alarms.

My core viewpoint is that the ideal way to prevent the risk of nuclear weapons use is: "Prevention is better than cure". In other words, we should prevent nuclear risks from occurring in the first place, and the next best thing is to control the risks. In my view, preventing the use of nuclear weapons should address the root causes, and practical, effective measures should be taken to eliminate the risk before it occurs.

References

1. L. Clavarino, Italian Physicists and the Bomb: Edoardo Amaldi's network for arms control and peace during the cold war. J. Contemp. Hist. **56**(3), 665–692
2. J.R. Biden, Why American must lead again: rescuing US. Foreign policy after Trump. Foreign Affairs **99**(2), 64 (2020)
3. F. Calogero, The time factor, in *Inadvertent Nuclear War: The Implications of the Changing Global Order*, 1st ed., ed. by H. Wiberg et al. (Pergamon Press, Oxford, 1993), p. 151
4. R. McNamara, What the U.S. can do. Newsweek **102**, 49–55 (1983)

Chapter 7
Public Opinion Between Rearmament and Crisis of the Nuclear Taboo

Fabrizio Battistelli and Francesca Farruggia

Abstract This article examines the changing relationship between the global public and nuclear weapons in the context of growing international insecurity. The authors describe the erosion of the concept of the "nuclear taboo", understood as a deep–seated normative aversion to nuclear weapons. By considering the standard surveys as abstract, the experimental surveys place respondents in simulated strategic scenarios designed to evoke more authentic responses. The difference between results—against nuclear weapons in standard surveys and favorable in experimental surveys—is explained by the fact that both have two different objects: standard surveys are about "rational" opinions and experimental ones about "emotional" attitudes, both useful to detect. In a standard survey conducted in Italy in 2025 reveals 81% of respondents consider nuclear weapons fundamentally different from conventional ones and always wrong to use. Among opponents the dichotomy between "deontologists" and "consequentialists" is discussed. The article concludes underlining the value of integrating affective and cognitive dimensions in understanding public opinion as a measure against the normalization of nuclear weapons in political discourse.

7.1 The Public's Aversion for Nuclear Weapons Between "Taboos" and "Tradition": What Methods to Study It?

Over the past fifteen years, a significant debate has emerged among scholars concerning nuclear weapons and how they are perceived by both governments and public opinion. In parallel with the deterioration of the arms control regime between the two nuclear superpowers—evidenced by the successive abandonment of treaties

---

F. Battistelli (✉)
Department of Social Sciences and Economics, University of Rome La Sapienza/Archivio Disarmo, Rome, Italy
e-mail: fabrizio.battistellli@uniroma1.it

F. Farruggia (✉)
Department of Social Sciences and Economics, University of Rome La Sapienza, Rome, Italy
e-mail: francesca.farruggia@uniroma1.it

P. Cotta-Ramusino et al. (eds.), *Nuclear Weapon Risks in a Problematic Time*, Springer Proceedings in Physics 358, https://doi.org/10.1007/978-3-032-29873-7_7

regulating the possession and deployment of the arsenals of the United States and Russia[1]—the nuclear threat has gained new momentum.

In Europe in particular, the invasion of a country that had renounced nuclear weapons (Ukraine) by a country possessing 5500 warheads (Russia) has generated growing concern among European public opinion. This has produced contradictory effects: on the one hand, renewed fears of military escalation; on the other, a resurgence of the relevance of nuclear weapons. For example, in European countries participating in the American "extended deterrence" strategy, such as Germany and the Netherlands, public opinion has shifted—from negative to positive—regarding the stationing of NATO missiles [16].

These are just some of the symptoms of a progressive erosion—partly spontaneous, partly deliberate—of the phenomenon known as the "nuclear taboo." According to this concept, popularized by Tannenwald [27–29], the particular characteristics of atomic technology and its ultimate product, the "Bomb," have generated among the population—and to some extent among elites—a widespread sense of aversion that has so far prevented its renewed use.[2]

Everything began in 1945 with the emergence on the strategic stage of a revolutionary weapon: the atomic bomb. What remains significant is the fact that, following the devastating conclusion of World War II, atomic technology has indeed proliferated to other states, but has never been used again—neither by the state that first employed it (the United States), nor by other states (primarily the Soviet Union) that adopted it in response.[3]

For an history's ironies, only two nuclear warheads have been used in the atomic era, while in the first seven decades alone, approximately 125,000 were produced [12]. When awarded the Nobel Prize in Economics in 2005, Thomas Schelling described the non-use of nuclear weapons since the two bombs dropped on Japan as

[1] The golden age of disarmament was the 15-year period that began in 1987 with the signing of the Treaty on the Elimination of Intermediate-Range and Shorter-Range Missiles (INF) and ended in 2002—on the one hand, with the signing of the Strategic Offensive Reductions Treaty (SORT) by the United States and Russia, and on the other, with both countries' withdrawal from the Anti-Ballistic Missile Treaty (ABMT). In 2007, Russia suspended its participation in the Treaty on Conventional Armed Forces in Europe (CFE), from which it formally withdrew in 2023. This withdrawal led to the treaty's earlier suspension and, in 2024, the official withdrawal of the United States. In rapid succession, the United Kingdom, France, Italy, Germany, Belgium, the Netherlands—as well as Belarus and Turkey—also withdrew from the CFE. In 2023, Russia additionally withdrew from the Comprehensive Nuclear-Test-Ban Treaty (CTBT). The START I treaties on strategic delivery systems expired in 2009, and SORT expired in 2011. These were replaced by the New START in 2010, which remains the last bilateral arms control agreement between Russia and the United States still in force, but it is set to expire in 2026. Furthermore, the United States withdrew from the Joint Comprehensive Plan of Action (JCPOA) on the Iranian nuclear program in 2018, and—crucially for Europe—from the INF Treaty in 2019.

[2] Independent and characterized by its own specificity, the so-called "renaissance" of civil nuclear power has re-emerged in several European countries—such as Italy, Germany, and the United Kingdom—which had previously appeared to abandon this technology [20].

[3] This marks a stark contrast with what had occurred over centuries with other revolutionary military innovations, such as the horse or gunpowder, which were rapidly deployed on the battlefield in their respective eras.

"astonishing" and referred to this fact as "the legacy of Hiroshima" [25]—the only positive aspect of an event universally regarded as tragic. It is worth noting, however, that following a long period of dominance by the nuclear taboo, in recent times the concept has been subject to progressive erosion, both spontaneous and deliberate. As a partial counterbalance to this trend, one can observe a renewed interest from the social sciences, particularly regarding public opinion. This has stimulated further study of the theme of "nuclear weapons use" on both a quantitative and qualitative level—on the one hand increasing the number of analyses, and on the other making them more sophisticated through the use of unconventional tools such as "experimental" surveys.

Within the lively scholarly debate on the existence and nature of the nuclear taboo, an alternative interpretation has gained ground: that the abstention from the use of such weapons is less a true taboo and more a tradition based "on the precedent and mutual adhesion models of reciprocal restraint" among international actors ([31]: 34; see also [17, 24]). Upon closer inspection, however, the definitional dichotomy between taboo and tradition is not so clear–cut—as well the debate about the application of a concept from cultural anthropology, like "taboo," appears nominalistic [15]. In our view, the conceptual opposition between "taboo" and "tradition" does not imply an irreconcilable incompatibility. What matters is the enduring aversion still held by a large majority—that nuclear weapons continue to provoke among the public across diverse social contexts. If we attribute to the term a heuristic rather than ontological value, it remains reasonable to use "taboo" as "shorthand for a strong normative use of nuclear weapons" ([2]: 166).

Regardless of adherence to one interpretation or another, the ongoing debate has nonetheless yielded some positive outcomes. On the one hand, it has revitalized a body of contributions that radically broaden the traditional perspective of law and international relations, encouraging social scientists to invest more research into issues—such as armaments—that had long been considered "exotic" in content and difficult to verify empirically. A leading role in this regard is played by one of the branches of the social sciences most advanced in its empirical approach to phenomena: public opinion research. In order to address the question of whether or not a taboo exists in a complex domain of human behaviour such as the use of a highly destructive weapon, the study of public opinion is driven to adopt innovative analytical perspectives, both conceptually and in terms of methodology and research techniques.

Before addressing conceptual aspects, we shall briefly consider the empirical ones, starting with the research associated with what has been termed "the second generation of taboo research" ([26]: 1072). The goal of this new approach is to examine assessments regarding public opinion (even more so than that of elites) on key aspects of the use of force, with particular reference to nuclear weapons. One of these assessments is the frequently cited notion of the United States' specificity as a country more inclined than its allies to consider the use of weapons in international crises. In his well–known metaphor, Robert Kagan claimed that: "Americans are from Mars and Europeans are from Venus" ([9]: 3). And to those, like Johan ([7]: 268), who observed that "to the man with a hammer the world looks like a nail", he replied that "nations without great military power face the opposite danger: when you don't have a hammer, you don't want anything to look like a nail" ([9]: 28).

In any case, according to the "second generation" of researchers, U.S. strategic policy has so far been studied thematically as a focus on the motivations and decisions of political elites and, methodologically, through a qualitative approach based on historical reconstructions of major events, case studies, interviews with key actors, and so on. Authors such as Sagan and Valentino propose an alternative to this approach, one that moves away from traditional opinion polls. They administer "experimental" surveys to panels of ordinary citizens, built around scenarios designed to simulate strategic emergencies as realistically as possible—contexts in which such individuals might actually have to take a position.

While we fully support the need to distinguish between the positions of decision–makers and those of ordinary citizens, we also note that the empirical study of the latter's opinions is not new. Following the events of September 11, 2001, and the subsequent "War on Terror," the use of military force prompted a significant surge of research and analysis focused on comparing American and European public opinion. For instance, regarding the 2003 war waged by President G.W. Bush against Saddam Hussein's Iraq—accused of possessing weapons of mass destruction and/or being complicit in jihadist terrorism—Pierangelo Isernia and Philip Everts, through their analysis of cross–national surveys, identified a "Transatlantic gap." Europeans differed not so much in terms of values (e.g., the democratic system and the need to defend it) from Americans, but rather in tactical preferences: placing "higher priority to soft tools" and considering the use of force as "a truly *ultima ratio* to be utilized only when all other sources of power have failed" ([8]: 255, see also [6]).

American "exceptionalism" is now being challenged by the second wave of research, particularly with regard to citizens' attitudes toward casualties suffered and inflicted in war—especially in the context of potential nuclear weapons use. The debate does not revolve around whether the American public is inclined to prioritize saving the lives of their own soldiers, even at the cost of disproportionately higher enemy casualties (combatants and non-combatants alike). In fact, "concerns about winning wars and the desire to minimize the loss of lives of their nation's soldiers dominate public opinion about military operations[4]" ([22]: 51). Rather, the emphasis is on the fact that such a stance is by no means exclusive to the American public, it is a shared feature of Western democracies, which tend to favour the use of all available military assets (including nuclear ones) that minimize the loss of their own uniformed personnel [23]. Through experimental surveys, researchers have assessed the willingness of publics in other nuclear–armed countries to support the use of such weapons in specific scenarios. Using the same methods, [4] surveyed not only samples of American citizens but also samples from the United Kingdom, France,

[4] Since the atomic bombings of Hiroshima and Nagasaki, the official justification for this dramatic decision has been based on the high estimated cost in human lives that the U.S. armed forces would have had to endure in the final phase of World War II, had they launched a ground invasion of Japan [27]. On the connected "taboo" of the immunity of non-combatants (lawed in the IV Geneva Convention, 1949) see [10].

and Israel—three officially nuclear states and one de facto nuclear state. The experimental nature of the survey lies in the presentation to each national panel of a (clearly fictional) newspaper article. The article describes a scenario in which a terrorist group based in Libya is preparing a chemical weapons attack against the four respondents' capitals (Washington, London, Paris, Tel Aviv).[5]

Experimental surveys have prompted various methodological and substantive critiques, to which we refer the reader to two dedicated debates published in *International Studies Review* [26] and *Security Studies* [18]. Here, we will address two of these critiques: (a) certain limitations in the inputs introduced into the surveys in order to make them "experimental", and (b) the stark discrepancy in respondents' willingness to support nuclear weapons use—relatively high in experimental surveys and significantly lower in standard ones. We begin by stating that, although we consider many of the critical observations directed at proponents of experimental surveys to be well–founded, we do not dismiss the underlying goal of these researchers: to find new ways of assessing and deepening our understanding of citizens' attitudes toward the use of nuclear weapons.

The first critique concerns the stimulation of respondents through inputs such as the fictional news article they are invited to read, and the degree of autonomy their responses can truly reflect within a "lab" context—one that is even more overtly artificial than standard survey settings. Of course, the use of stimuli such as cue cards, photographs, vignettes, or the presentation of documentary or custom–prepared texts is neither unknown nor unusual in social science research. However, it should be noted that such surveys typically address topics that are far less emotionally charged than those presented in the "mock news story," which includes casualty estimates for different types of attacks as well as highly influential details, such as the "opinion" of military leaders.[6]

On highly emotional topics such as those involving the antithetic couples of security/insecurity, aggression/defense (ultimately: life/death), it becomes difficult to control the affective dimension of participants involved in these experiments. In response to critics who argue that "pre–emptive scenarios" involving terrorists, North Korea, or Iran are "tough tests," Sagan and Valentino reply that they are, rather, "realistic tests." By introducing figures and circumstances related to plausible enemies of the United States, they are testing "the nuclear taboo in exactly the [political] conditions in which it might be needed to prevent nuclear use" (Sagan and Valentino 2021:

[5] The dilemma posed concerns the means that each government should employ to neutralize the threat: whether to strike the terrorists using relatively less effective conventional weapons (with a 45% chance of destroying the target) or to resort to more decisive nuclear weapons (with a 90% chance of success). This choice is presented under the condition that both options would cause the same number of casualties (2700), to be weighed against the estimated 3000 victims if the attack were to occur [4]. The data show that "American attitudes toward nuclear weapons and on combatant immunity are not exceptional [...] contrary to many scholars expectations, the publics of two other nuclear-armed democracies are equally willing (France) as or more willing (Israel) than Americans to use nuclear weapons and deliberately target civilians" (Press et al. 2022: 28).

[6] See the four-column headline: "Terrorists Planning Chemical Weapons Attack on Washington, D.C. [Paris, London, Tel Aviv]. Joint Chiefs Say Nuclear Airstrike Doubles Chances of Destroying Terrorist Chemical Lab in Libya," Appendix 1: News Stories (U.S. versus) in [4].

1093). Rejecting the notion that their respondents are "too emotionally invested," the authors argue that, on the contrary, they may not be emotionally invested enough ([5]: 197). Indeed, the experiments are quite far from "the depth of urgency and emotion that many Americans felt in 1945" and constitute "a conservative test of the public's willingness to use nuclear weapons and target members of non-combatants in war" ([22]: 44).

Undeniably, manifestations that are not entirely rational are far from marginal aspects of the human condition and cannot be ignored. They are, in fact, integral to the worldview internalized by respondents—especially in contexts such as those described in experimental scenarios, which approximate a highly emotional reality. As current events in this third decade of the twenty-first century show, far from being dismissed as irrelevant or discredited as relics of the past, irrational elements are increasingly emerging as central actors in both public expressions and the behavior of decision–makers. For some time now, even the natural sciences have moved toward accepting—something the social sciences were the first to discover—that scientific knowledge is necessarily reductionist, destined to carve out limited segments of reality. In fulfilling its purpose, it is bound to reduce complexity at the very moment it seeks to impose cognitive order on what Max Weber called the "senseless infinity" [32]. The effort of scientific knowledge to recreate in vitro conditions that may represent even a distant proxy of historical reality cannot be dismissed, provided that researchers remain aware of the approximate nature of their endeavor.

Even more challenging is the second critique. Benoît Pelopidas and Kjølv Egeland argue that the findings presented in *Kettles of Hawks*—the largest of the experimental surveys conducted by the group [4]—"contrast with several existing polls and surveys that indicate relatively strong public opposition to nuclear use" in France, the United Kingdom, and the United States itself ([18]: 190).

In our view, the explanation for this clear discrepancy does not lie primarily in differences of style, wording, or scenarios—although these certainly exist and undoubtedly contribute to distinguishing standard surveys from experimental ones. Rather, *the discrepancy stems from the difference in the object being measured*. The entire contrast between standard and experimental surveys can be resolved if we acknowledge that both are cognitively directed at two distinct, albeit related, phenomena. This is made possible through the valuable recovery of the distinction between the two concepts of opinion and attitude. The overlap between these two basic modes of human cognition had already drawn the attention of one of sociology's foremost figures, Robert K. Merton. In his theoretical interpretation of the *American Soldier* research campaign, Merton warned of the risk that "basic conceptual differences [...] might be obscured by a single crudely defined concept," such as "the single blanket concept of 'attitude'" ([13]: 73). Unlike *opinion*, which captures aspects of reality at a high level of abstraction, *attitude* reflects more practical and immediate aspects. Aware of the limited influence that political opinions had on soldiers' performance of their duties, the researchers led by Samuel A. Stouffer focused instead on "sentiments and attitudes" (e.g., "the willingness for combat") ([13]: 45).

Although the prevailing trend over time has been, improperly, to use the terms "attitude" and "opinion" interchangeably, the conceptual difference between them becomes even clearer when viewed through disciplinary lenses: as Bergman [3] observes, social psychologists have typically favored the concept of attitude, political scientists that of opinion, and social anthropologists that of values. The usefulness of employing the concept of attitude as a predominantly affective and emotional manifestation has been reaffirmed, among others, by Aiken [1], as distinct from the concept of opinion, which expresses explicit and conscious positions—what we might cautiously call "rational."

Applied to our topic, the controversy between standard and experimental surveys can be reconciled by explicitly recognizing the distinct objects targeted by each research tool: on the one hand, standard surveys aim to capture a detached and "rational" representation (albeit limited by abstraction); on the other, experimental surveys aim to capture an involved and "emotional" representation (albeit shaped by the constraints of the laboratory setting).

7.2 A Survey: Opposition to the Use of Nuclear Weapons Between "Deontologists" and "Consequentialists"

On the topic of public opposition to—or acceptance of—the use of nuclear weapons, the distinction between those who reject a priori such a possibility (*deontologists*) and those who base their judgment on an a posteriori evaluation of the expected (positive or negative) consequences (*consequentialists*) proves especially useful [4]. The crucial role of consequences in political decision-making—and thus the centrality of accepting responsibility for them—is fundamental in classical liberal thought, and particularly in Weberian sociology.

Hence, the unavoidable dilemma that, according to Weber [33], accompanies most decisions in the political sphere: between the ethics of conviction (based on adherence to values) and the ethics of responsibility (based on consideration of consequences).

To assess the nature and distribution of opinions (in the narrow sense described in the previous paragraph), we conducted a standard survey on a sample of 802 Italian citizens aged 18 and older, representative by gender, age, and geographic residence in a country like Italy, which (at least as of spring 2025) participates in the American extended deterrence framework.

Beginning with perceptions of whether nuclear weapons are distinct from conventional weapons, 81% of respondents consider them fundamentally different and believe their use is always wrong. In contrast, 10% view nuclear weapons as merely more powerful than conventional ones and find their use appropriate in certain cases. Finally, 9% responded with "don't know/no answer" (Fig. 7.1).

Using a hierarchical binary logistic regression model, we first assessed the impact of sociodemographic variables on how respondents distinguish between conventional and nuclear weapons. Taking into account gender, age, education level, and political orientation, only age group and political orientation were found to be statistically significant. With respect to age, the younger the respondent, the more likely they are

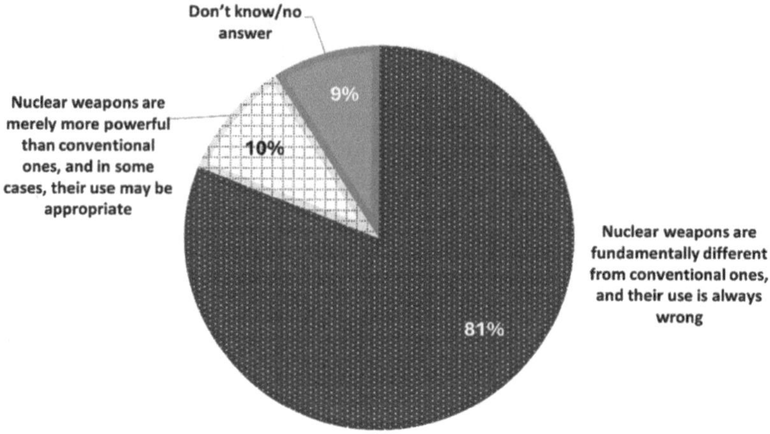

Fig. 7.1 When it comes to conventional and nuclear weapons, which of the following two statements do you agree with? *Source* Archivio Disarmo, Difebarometro No. 11, February 2025. CAWI opinion survey on a sample of 802 respondents representative of the Italian population, January 2025

to consider nuclear weapons as merely more powerful than conventional ones and, in certain cases, their use as appropriate. As for political orientation, those identifying as right-wing or center-right are more inclined—compared to those on the left—to view the use of nuclear weapons as situationally justified.

We also asked respondents to answer two questions reflecting, respectively, an attitude and an opinion. The attitude involved a hypothetical situation: if one were the victim of a crime, would they turn to the police or take matters into their own hands, even using a weapon? The opinion concerns support or opposition to the introduction of the death penalty for the most serious crimes, such as mass murder or massacres. Two groups emerged: on one side, *reactive* respondents, who support harsh punishments and self-execution of justice; on the other, *adaptive* respondents, who support proportionate punishments administered by the State. The category of reactive respondents was subsequently added as an independent variable in the hierarchical binary logistic regression model.

Compared to non-reactive group, reactive respondents are statistically more inclined to believe that nuclear weapons are merely more powerful than conventional ones and that their use may be appropriate in certain cases. With the inclusion of the "reactive respondents" variable, age continues to show a statistically significant correlation with how respondents evaluate nuclear versus conventional weapons, while political orientation loses statistical significance. This change is due to the presence of a statistically significant association between political orientation and reactive respondents: those who identify as right-wing or center-right, rather than left-wing, are more likely to fall into the reactive group.

Subsequently, we asked the 81% of respondents (653 out of 804) who considered nuclear weapons fundamentally different from conventional ones and always wrong to specify the reasons for their position. The aim was to explore the broad range of responses which, as shown in the literature and in our previous research, reflect the variety of reasons underlying the nuclear taboo and the closely related issue of civilian immunity.

In order to systematize the range of responses into a conceptually structured framework, we identified three groups, arranged along the following axes: on the x-axis, the poles A priori/A posteriori; and on the y-axis, the poles Particularism/Universalism. Regarding the first axis, in the A priori category, opposition expressed by respondents is based on ethical and legal reasons; like the Kantian categorical imperative of "you must," a choice (in this case, the rejection of nuclear weapons) is made regardless of its costs or consequences, simply because it is morally right (hence the label "*Deontologists*" for those who hold this position).[7]

Conversely, in the a posteriori category, opposition is based pragmatically on a rational cost–benefit analysis focused on the expected consequences of a given decision (hence the term consequentialists for those who share this second position). Consequentialists, in turn, can be divided into two subgroups according to the Particularism/Universalism axis. While those who are unconditionally (a priori) opposed to nuclear weapons—presumably based on a universalistic principle—a posteriori opponents reject nuclear weapons for two distinct types of reasons. On one side are motivations that may be defined as *particularistic consequentialist*: they express opposition to nuclear weapons, but based on the harm that could befall the user, either through enemy retaliation or by creating a dangerous precedent that may backfire in the future. On the other side are motivations also centred on harmful consequences, but characterized by a broader perspective: harm done not only to oneself, but also to others. We define these as *universalistic consequentialist* motivations. They do not limit concern to foreseeable harm affecting oneself, one's family group, or nation, but extend it to the predictable harm suffered by all civilian populations (friendly, neutral, and even enemy), as well as by the environment.

This conceptual classification of motivations for opposition to nuclear weapons use is presented in Fig. 7.2.

By applying the three categories of motivation to the 81% of respondents who consider nuclear weapons fundamentally different from conventional ones and believe their use is always wrong, we identified three corresponding groups. The largest of these (59%) consists of *universalistic consequentialists*. This group includes 51% who base their judgment on the serious and lasting harm nuclear weapons can inflict on innocent civilians, plus 8% who cite the environmental damage such weapons can cause.

In second place, at 23%, are the *particularistic consequentialists*: 15% are concerned about possible nuclear retaliation by one or more nuclear-armed states, and 8% are concerned about the precedent that might be set for states that possess—or could come to possess—nuclear weapons in the future.

[7] The ethical dilemma between a deontological choice and a consequentialist choice—illustrated by the well-known trolley problem—is extensively addressed in numerous studies of moral philosophy.

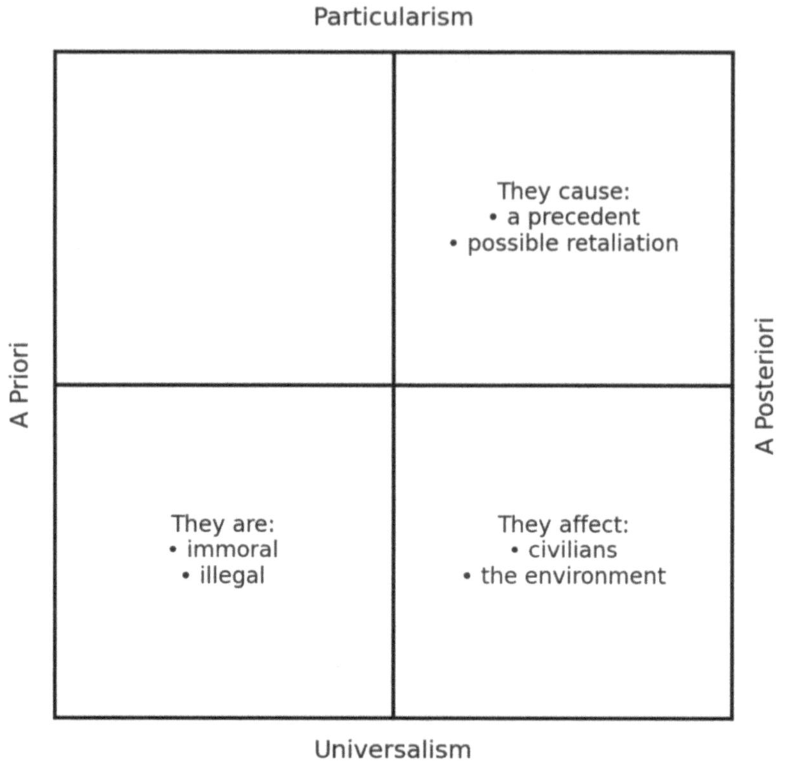

Fig. 7.2 Classification of motivations for rejecting the use of nuclear weapons

Finally, 18% are deontologists (who, by definition, are universalists), believing nuclear weapons are always wrong due to their inherently immoral nature (12%) or their violation of international law (6%) (Fig. 7.3).

Using a multinomial logistic regression model in which the most common category *universalistic consequentialists*—was set as the reference group, we evaluated the impact of sociodemographic variables, as well as whether or not respondents belong to the category of reactive individuals, on the motivations underlying a negative assessment of the consequences of using nuclear weapons.

A first finding is the statistically significant impact of gender when comparing *universalistic consequentialists* with both *deontologists* and *particularistic consequentialists*. Specifically, the analysis shows that being male increases the likelihood of being a *deontologist* rather than a *universalistic consequentialist*. In the comparison between *particularistic* and *universalistic consequentialists*, both gender and reactivity play a role: being male increases the likelihood of adopting a particularistic stance, and being a reactive individual significantly increases the probability of being particularistic rather than universalistic. However, no statistically significant difference was found between *deontologists* and *universalistic consequentialists*.

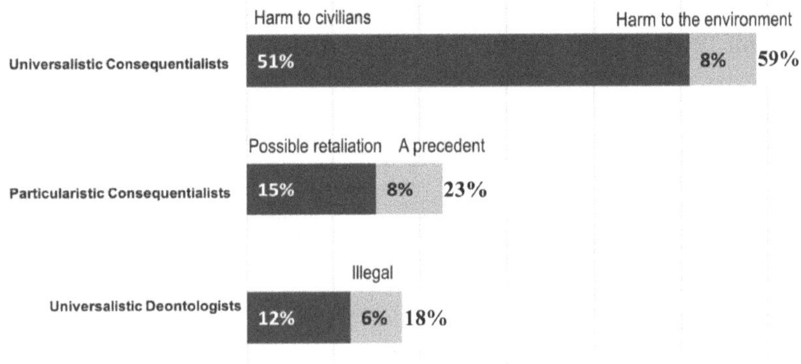

Fig. 7.3 Breakdown of motivations for rejecting the use of nuclear weapons. *Source* Archivio Disarmo, Difebarometro No. 11, February 2025, sub-sample of 653 respondents who consider the use of nuclear weapons always wrong

Shifting from the general evaluation of nuclear versus conventional weapons to an analysis of the motivations for rejecting the use of nuclear weapons, we observe that gender—while not decisive in evaluating their use per se—has a statistically significant impact on why some respondents reject nuclear weapons. Women, more attuned to the practical consequences of their use rather than opposing them a priori, tend to adopt a universalistic consequentialist perspective more often than men, who are more likely to lean toward particularistic consequentialism.

Age, by contrast, does not appear to influence the reasons for rejecting nuclear weapons, but it does affect the general assessment of their use: as age increases, the likelihood of supporting the use of nuclear weapons decreases. Finally, being a reactive individual—defined by approval of the death penalty and armed self-defense—has an impact on both the evaluation of nuclear weapons use and the motivations associated with it. On the one hand, reactive individuals are more inclined to support the use of nuclear weapons in certain situations; on the other hand, among those who always oppose their use, reactivity is associated with particularistic consequentialist reasoning, based on fear of the potential harm their use could cause.

7.3 Conclusion

In a domain as highly complex as the one addressed in these notes, it is more appropriate to speak of approximations to a multidimensional reality than of definitive conclusions. Few issues present, as does that of nuclear armaments, such a complex intersection of variables—each one difficult to reduce to simple entities and, on the contrary, serving as a nexus for crucial physical, technological, psychological, social, economic, political, and ideological dimensions—just to name the main ones.

Each of these dimensions contributes to defining the issue of nuclear weapons and, more specifically for our purposes, their role in the eyes of public opinion. In our analysis, we have touched on the question of whether the term "taboo" is appropriate when referring to nuclear weapons. Should this term be considered excessive or not entirely suitable, it may be replaced with the less evocative and more descriptive term "aversion." The result does not alter the substance of the matter. In recent years, alongside the deterioration of international relations, there has been an effort—driven by politics and mainstream media—to "normalize" both the language and content associated with nuclear weapons. Despite undeniable progress in this direction, all national and cross-national public opinion surveys continue to show that this technology and strategic asset still inspire concern and hostility among the majority of the public in virtually every country in the world.

Whether this inhibition is vaguely sacred (a "taboo") or more secular (a "tradition") the use of nuclear weapons is still regarded, even by the elites of nuclear-armed countries, as a last resort: one that is heavily relied upon virtually (in terms of threats), but from which actual use has so far been abstained.

At the same time, no one can guarantee that what currently appears to be a reasonable expectation of no first use (supported by taboo or aversion—either way, by the "legacy of Hiroshima") will endure indefinitely. In the delicate balance whereby governing elites are, to some extent, influenced by public opinion, and citizens are influenced by nuclear weapons in the sense of opposing them, various and often contrasting factors interact in shaping the positions taken by both rulers and the ruled. On one side are rational factors (more or less aligned with a particular morality), and on the other, irrational factors (also converging, more or less, toward a different morality).

Our interpretation is that both types give rise to two distinct "products": reasoning manifests in opinions, while affectivity and emotion are expressed through attitudes. Since the human actor is an expression of both rational and affective-emotional components, both opinions and attitudes deserve recognition within individuals and groups, and both require scholarly investigation. It is therefore legitimate for different disciplinary and theoretical perspectives to choose which object to focus on. What matters most is that they do so with full awareness of their own epistemological and methodological strengths and limitations—and above all, that they succeed in bringing their findings to the attention of citizens and decision-makers, in pursuit of the shared objective of preventing nuclear war.

References

1. L.R. Aiken, Attitudes and related psychosocial constructs, in *Theories, Assessment and Research* (Sage Publications, Thousand Oaks, 2002)
2. M.S. Bell, The nuclear taboo and the inevitability of uncertainty. Secur. Stud. **32**(1), 166–204 (2023). https://doi.org/10.1080/09636412.2023.2178966
3. M.M. Bergman, A theoretical note on the differences between attitudes, opinions, and values. Swiss Polit. Sci. Rev. **4**(4), 81–93 (1998). https://doi.org/10.1002/j.1662-6370.1998.tb00239.x

4. J. Dill, S.D. Sagan, B.A. Valentino, Kettles of Hawks: public opinion on the nuclear taboo and noncombatant immunity in the United States, United Kingdom, France and Israel. Secur. Stud. **31**(1), 1–31 (2022). https://doi.org/10.1080/09636412.2022.2038663
5. J. Dill, S.D. Sagan, B.A. Valentino, Public opinion and the nuclear taboo across nations: an exchange—the authors reply. Secur. Stud. **32**(1), 195–204 (2023). https://doi.org/10.1080/09636412.2023.2178971
6. P. Everts, P. Isernia (eds.), *Public Opinion and the International use of Force* (Routledge, London, 2001)
7. J. Galtung, Peace by peaceful means, in *Peace and Conflict, Development and Civilization* (Prio, Oslo, 1996)
8. P. Isernia, P. Everts, Partners apart? The foreign policy attitudes of the American and European publics. Japanese J. Polit. Sci. **5**(2), 229–258 (2004). https://doi.org/10.1017/S1468109904001501
9. R. Kagan, *Paradise and Power: America and Europe in the New World Order* (Vintage Books-Random House, New York, 2003)
10. H.M. Kinsella, *The Image Before the Weapon: A Critical History of the Distinction between Combatant and civilian* (Cornell University Press, Ithaca, NY, 2011)
11. L.L. Koch, M. Wells, Still taboo? Citizens' attitudes toward the use of nuclear weapons. J. Glob. Secur. Studies **6**(3), 1–18 (2021). https://doi.org/10.1093/jogss/ogaa024
12. H.M. Kristensen, R.S. Norris, Global nuclear weapons inventories, 1945–2013. Bull. Atomic Scientists **69**(5), 5–81 (2013). https://doi.org/10.1177/0096340213501363
13. R.K. Merton, A. Kitt Rossi, Contributions to the theory of reference group behaviour, in *Continuities in Social Research*, ed. by R.K. Merton, P.F. Lazarsfeld (The Free Press, New York, 1950), pp. 40–105
14. A.H. Montgomery, C. Carpenter, Whence the nuclear taboo?, in *Forum: Nonuse of Nuclear Weapons in World Politics; Toward the Third Generation of 'Nuclear Taboo' Research*, ed. by M. Smetana, C. Wunderlich. Int. Stud. Rev. **23**(3), 1089–1092 (2021)
15. H. Müller, Taboo or tradition or what? A critical look at the terminology and conceptualization of nuclear venues, in *Forum: Nonuse of Nuclear Weapons in World Politics; Toward the Third Generation of 'Nuclear Taboo' Research*, ed. by M. Smetana, C. Wunderlich. Int. Stud. Rev. **23**(3), 1082–1085 (2021)
16. M. Onderco, M. Smetana, T.W. Etienne, Hawks in the making? European public views on nuclear weapons post-Ukraine. Glob. Policy **14**(2), 305–317 (2023). https://doi.org/10.1111/1758-5899.13179
17. T.V. Paul, *The Tradition of Non-use of Nuclear Weapons* (Stanford University Press, Stanford, CA, 2009)
18. B. Pelopidas, K. Egeland, Thinking about what people think about nuclear weapons. Secur. Stud. **32**(1), 188–194 (2013). https://doi.org/10.1080/09636412.2023.2178969
19. D.G. Press, S.D. Sagan, B.A. Valentino, Atomic aversion: experimental evidence on taboos, traditions, and the non-use of nuclear weapons. Am. Polit. Sci. Rev., 1–19 (2013). https://doi.org/10.1017/S0003055412000597
20. B. Renzi, M.D. Cotton, G. Napolitano, R. Barkemeyer, Rebirth, devastation and sickness: analyzing the role of metaphor in media discourses of nuclear power. Environ. Commun. **11**(5), 624–640 (2017). https://doi.org/10.1080/17524032.2016.1157506
21. M.R. Rublee, The logic of affect and nuclear taboo research, in *Forum: Nonuse of Nuclear Weapons in World Politics; Toward the Third Generation of 'Nuclear Taboo' Research*, ed. by M. Smetana, C. Wunderlich. Int. Stud. Rev. **23**(3), 14–18 (2021)
22. S.D. Sagan, B.A. Valentino, Revisiting Hiroshima in Iran. What Americans really think about using nuclear weapons and killing noncombatants. Int. Secur. **42**(1), 41–79 (2017). https://doi.org/10.1162/ISEC_a_00284
23. S.D. Sagan, B.A. Valentino, Not just a war theory. Int. Stud. Q. **62**(3), 929–937 (2018). https://doi.org/10.1093/isq/sqy033
24. S.D. Sagan, Realist perspectives on ethical norms and weapons of mass destruction, in *Ethics and Weapons of Mass Destruction: Religious and Secular Perspectives*, ed. by S. Hashmi, S. Lee (Cambridge University Press, Cambridge, 2004), pp. 73–95

25. T.C. Schelling, An astonishing sixty years: the legacy of Hiroshima. Am. Econ. Rev. **96**(4), 929–937 (2006). https://doi.org/10.1257/aer.96.4.929
26. M. Smetana, C. Wunderlich, Forum: nonuse of nuclear weapons in world politics: toward the third generation of "nuclear taboo" research. Int. Stud. Rev. **23**(4), 1072–1099 (2021). https://doi.org/10.1093/isr/viab002
27. N. Tannenwald, The nuclear taboo: the United States and the normative basis of nuclear non-use. Int. Org. **53**(3), 433–468 (1999). https://doi.org/10.1162/002081899550959
28. N. Tannenwald, *The Nuclear Taboo: The United States and the Non-Use of Nuclear Weapons Since 1945* (Cambridge University Press, Cambridge, 2007)
29. N. Tannenwald, How strong is the nuclear taboo today? Washington Q. **41**(3), 89–109 (2018). https://doi.org/10.1080/0163660X.2018.1520553
30. N. Tannenwald, Public support for using nuclear weapons on Muslims: a response to Sagan, Valentino, in *Forum: Nonuse of Nuclear Weapons in World Politics; Toward the Third Generation of 'Nuclear Taboo' Research*, ed. by M. Smetana, C. Wunderlich. Int. Stud. Rev. **23**(3), 1078–1082 (2021)
31. T. Ward, *The Ethics of Destruction: Norms and Force in International Relations* (Cornell University Press, Ithaca, NY, 2001)
32. M. Weber, The objectivity of knowledge in social science and social policy, in *Max Weber: Collected Methodological Writings*, ed. by H.H. Brun, S. Whimster (Routledge, Abingdon-on-Thames, 1904), pp. 100–138
33. M. Weber, Politics as a vocation, in *Max Weber: the Vocation Lectures, Edited and with an Introduction*, ed. by D. Owen, T. B. Strong (Hackett Publication, Indianapolis, 1919), pp. 32–94

Part II

Chapter 8
The Dark Side of ICT, Cyber-Info Warfare and International Security

Gian Piero Siroli

Abstract This paper explores the significant and evolving impact of Information and Communication Technologies (ICT), particularly Artificial Intelligence (AI), on conventional warfare, international security and the nature of conflict itself; it also highlights the intrinsic dual-use nature of ICT and the increasing digitization of the battlefield. The integration of AI is discussed in the context of military operations, including its role as a force multiplier and in the development of Lethal Autonomous Weapon Systems (LAWS). The paper examines the proliferation and capabilities of various unmanned systems (aerial, ground, marine) and the associated trends, challenges and ethical considerations, including the principle of Meaningful Human Control (MHC). It delves into the potential impact of AI on nuclear weapons and global strategic stability, identifying risks of accidental escalation and calling for transparency and limitations on AI use in nuclear forces. Intrinsic vulnerabilities of AI, such as adversarial attacks, opacity and data poisoning, are detailed. The landscape of cyber-info warfare is explored, focusing on information and psychological operations, the use of social media for disinformation and the challenge of deepfakes. Finally, the paper reviews international efforts and discussions within the United Nations and other for a concerning ICT security, the application of international law and the urgent need for regulation regarding autonomous weapons and the broader risks posed by AI to global peace and security. The pervasive influence of AI technologies raises fundamental questions about control, trust and the potential opening of a new ethical "Pandora's box" similar to the development of nuclear weapons in the past.

G. P. Siroli (✉)
Department of Physics and Astronomy, University of Bologna, INFN & CERN, Bologna, Italy
e-mail: Gian.Piero.Siroli@cern.ch

P. Cotta-Ramusino et al. (eds.), *Nuclear Weapon Risks in a Problematic Time*, Springer Proceedings in Physics 358, https://doi.org/10.1007/978-3-032-29873-7_8

8.1 Introduction: The Dual Nature of ICT and Warfare Digitization

ICTs are fundamentally dual-use technologies, meaning they have both beneficial civilian applications and potentially harmful military applications. The increasing integration of ICT is driving the digitization of conventional warfare domains of land, sea, air and space. The current battlefield digitization is incorporating the full spectrum of ICTs, including Cloud Computing and Big Data, Artificial Intelligence, Machine Learning (ML), networking technologies and cyber-security. A significant consequence of this fast-developing process is the observation that above a certain speed of machine autonomous operations, humans are overcome and overwhelmed.

8.2 Artificial Intelligence in Military Operations

AI is being integrated across the full spectrum of military operations. Its applications range from increasing the reliability and efficiency of defense infrastructures in real time, particularly against AI-enabled cyber threats on critical networks, through pattern recognition and anomaly detection in various domains; AI is also employed to process and manage sensor data, monitor systems integrity and support vocal commands. AI serves as a force multiplier in digital spaces, addressing challenges of scale and sophistication in areas like threat monitoring and response, navigation and it is also applied in offensive and defensive cyber operations.

Beyond support functions, AI is deeply involved in human decision-making support and in the control of (lethal) autonomous weapon systems, which are described as being in a "VERY advanced prototype/test phase". There is an ongoing international debate about the level of human involvement needed in LAWS, leading to the crucial concept of "Meaningful Human Control" (MHC), which is deemed necessary to maintain oversight and accountability in compliance with ethical, moral and legal constraints.

Current algorithms are not (yet?) capable of human-level reasoning, but we already have weapons that can use AI to autonomously search, select and engage targets in specific situations, so it is considered essential for humans to retain the ultimate decision on life and death, and not delegate it to a machine.

8.3 Unmanned Systems: Evolution and Impact

Unmanned systems, including vehicles used for various military operations with different levels of autonomy (from remotely controlled to fully autonomous)[1] and equipped with advanced sensors, data processing capabilities and communication technologies, are in an evolutionary phase of integration in different warfare domains:

- Aerial Vehicles (UAVs): Used for long-range stealth missions such as surveillance and reconnaissance and target acquisition, but also at the operational level for combat support and electronic warfare. They can share data in real time with manned aircrafts and cooperate with pilots, as in the case of Collaborative Combat Aircrafts (CCAs) (e.g., XQ-58A Valkyrie).
- Ground Vehicles (UGVs): Land robotic systems primarily used for reconnaissance, reducing human risk in hazardous environments, bomb disposal and logistics. They feature autonomous navigation, obstacle avoidance and payload delivery (e.g., M5 UGV).
- Marine Systems (UMVs): Include underwater (UUV) drones and surface (USV) vessels for coastal surveillance, anti-submarine warfare and mines countermeasures. They are capable of autonomous operation and long-duration missions for data collection (e.g., Orca XLUUV).

Key trends in unmanned systems include increased autonomy, rapid integration of AI and military clouds for data collection/distribution/analysis, enhanced interoperability between unmanned and manned systems, improved human–machine interfaces and the development of drone swarms.

These trends bring significant challenges, including cybersecurity vulnerabilities, legal issues (compliance with international laws of armed conflict) and ethical considerations. Ethical challenges specifically relate to autonomous decision-making in combat scenarios, particularly regarding the use of lethal force, and the need for clear lines of accountability for actions taken by these systems in complex operational environments. The principle of Meaningful Human Control, previously mentioned, is central to these discussions.

Unmanned autonomous systems and drones are indeed reshaping military strategies and tactics. This involves the integration of cyber-physical military assets with traditional warfare domains and its incorporation into existing organization structures, enhancing Intelligence, Surveillance, Reconnaissance (ISR) and situational awareness, facilitating electronic warfare, improving targeting and precision strikes, offering cost-effectiveness, increasing operational safety (by keeping personnel out of harm's way) and influencing asymmetric warfare. In this context, the US Department of Defense outlined its vision in the document "Unmanned Systems Integrated Roadmap 2017–2042".

[1] Described by the concepts of "Human in, on, or out of the loop" (remote, semi-autonomous, autonomous control system).

8.4 AI and Nuclear Risk

Advances in AI could potentially have a significant impact on the field of nuclear weapons and posture, with consequences for strategic stability and nuclear risk. Machine Learning and autonomy could unlock new possibilities for a wide array of nuclear force-related capabilities, such as early warning, command and control and weapon delivery. There is clear evidence that all nuclear-armed states have prioritized the pursuit of AI, raising questions about the stabilizing or destabilizing effects of such developments.

One concern is that investment in AI by an adversary, even in non-nuclear-related areas, could threaten a state's future second-strike capability, the base of current nuclear deterrence theory, potentially generating insecurity, decreasing strategic stability and increasing the risk of nuclear conflict. Furthermore, AI could fail or be misused in ways that could trigger an accidental or inadvertent escalation of a crisis or a conflict into a nuclear confrontation.

Addressing these risks requires specific measures:

- Support awareness-raising among stakeholders (governments, industry, civil society) about the challenges posed by AI in the nuclear arena.
- Support transparency and Confidence-Building Measures (CBMs) to reduce misperception and misunderstanding among nuclear-armed states.
- Discuss and agree on concrete limits to the use of AI in nuclear forces and infrastructures.

8.5 AI Intrinsic Vulnerabilities and Challenges

The increasing pervasiveness of AI gives rise to the risk of so-called "adversary AI" which exploits weaknesses in machine learning models to misinterpret inputs and behave favorably to an attacker. Attackers can create "adversarial examples", malicious inputs that appear as normal but are meticulously crafted and optimized to break the models, leading to instability and inaccurate predictions. This can involve exploiting specific behaviors within AI internals (such as neural networks), which may be unknown to the developers. The current opacity of AI internals means they generally function as a "black box": even designers may not be able to explain why an AI system reaches a specific result. This opacity hinders interpretability, predictability, verifiability and reliability of the overall AI system.

Other vulnerabilities include:

- "Poisoning attacks" during the supervised learning phase, where wrong, noisy, manipulated or non-balanced data may be introduced, leading to "Garbage In/ Garbage Out effect"). This can also cause biassed output at the intellectual or ethical level.

- Neural Networks can also be used to hide, store or transfer malicious payloads and/or trigger malware. Techniques exist for inserting imperceptible and undetectable backdoors into compiled neural networks.
- Many different classes of failures, intentional and unintentional, have been identified in Machine Learning techniques.

Considering all those vulnerabilities and risks, one can wonder whether AI stands for Artificial Intelligence or "Automating Ignorance". There is a recognized need for an "explainable" Artificial Intelligence (XAI) to ensure that AI results can be understood by humans. XAI is currently an active research field.

Many reports and documents address the issue of trustworthy and responsible AI. Care has to be taken when productionizing reliable AI systems: the full production chain must be secured, including algorithms, training processes and data, management of external dependencies. From the point of view of risk management, failure modes also must be considered, like model input, performance bias and robustness failures.

Recently the UN Secretary-General A. Guterres recognized that "AI has tremendous potential but also major risks" noting that "even its own designers have no idea where their stunning technological breakthrough may lead".

The vulnerabilities previously described are at the same time also challenges for AI future development:

- The AI training phase needs accurate, very reliable, high quality, unbiased datasets, not easy to collect or create.
- Human supervision and feedback needed to fine-tune AI models, can be time-consuming, costly, subjective and inconsistent.
- The current "black box" approach should be overcome to reach model transparency and interpretability.
- Adversarial attacks should be brought under control or at least to be detectable, to increase overall robustness and safety.
- Hallucinations: AI models sometimes generate false or misleading output/information that appears plausible but is entirely incorrect, illogical or out of context. Factors like biased datasets, incomplete information or overfitting[2] can contribute to this phenomenon which strongly undermines AI reliability and trustworthiness. Beyond hallucinations, some sophisticated *prototypes* of AI Language Models *appear* to exhibit skills that were not explicitly planned or programmed, as if they were out of human control.
- Slow time-drifting conditions in unsupervised learning can cause the AI models to struggle in adapting to evolving patterns, causing a decline in accuracy and reliability and leading to model failures.

Unfortunately, the potential for misuse is significant: relatively recently AI models have been repurposed to design toxic or deadly biochemical weapons, with strong implications for the Chemical and Biological Weapons Conventions.

---

[2] "Overfitting" in AI happens when a model learns too much detail from its training data to the point where it struggles to generalize to new, unseen data.

It is universally recognized the need for "ethical filters", mechanisms to ensure that AI systems operate within ethical boundaries. These filters are crucial for responsible AI development, however at the time of writing, ethical filters can be somehow bypassed or manipulated.

The previous points may raise the following questions: is all this safe? Can we control this technology? How can we trust AI? A lot of research work will be needed to face these challenges.

8.6 Cyber Info Warfare and Information Operations

Considering the Internet as a global communication medium, rather than just bits, bytes and protocols, reveals its role in the military context of Information Operations (IO). IO involves information manipulation for purposes such as propaganda and counter-propaganda, disinformation, consensus building, de-legitimation, censorship and content filtering. Information Operations, also called Psychological Operations (PSYOPS), allow for deception and influence on perceptions, emotions, values and behaviors of target individuals or groups, also in socio-political contexts. These are all traditional, centuries old, techniques but applied on a new medium of communication with its specific characteristics, in particular the fast propagation speed and distribution at the planetary level. Real-world examples include support for dissident groups, recruitment campaigns, large-scale disinformation via social media, and notable events like the EZLN's "Social Netwar," Wikileaks, NSALeaks and CIALeaks cases and Cambridge Analytica scandal.

The digital network environment can be described as an ubiquitous surveillance environment. Information warfare has a primary political and strategic value and can affect the political and social instability of a country. This view also blurs the distinction between military and civilian domains and infrastructures.

Many countries have special units dedicated to PSYOPS. Information operations take place not only during wartime but even during peacetime, clearly showing that the military front is not the only battlefield.

AI can play a very significant role in PSYOPS by enhancing the precision and effectiveness of influence campaigns. It helps military and strategic actors analyze vast amounts of data from social media, tailor messaging to profiled classes of users/ targets, predict responses and optimize engagement strategies. In this context of polarized and deceptive communication AI-based fake news, including images, audio and video deep-fakes (created by AI and *currently* undetectable by AI) turn out extremely useful tools.

The overall impact is raising a side question: "Is AI destroying credibility of, and on, the Internet?"

8.7 International Security and Governance

The international community has been grappling with the security of ICTs in the context of international security for many years. The UN is having a leading role in promoting dialogue, developing common understandings on international law application, and norms for responsible State behavior. Annual reports and resolutions by the Secretary-General to the General Assembly on "Developments in the field of information and telecommunications in the context of international security" date back to 1998.

Multiple Groups of Governmental Experts (GGEs), with limited participation to about 20 UN member states, have examined existing and potential threats from the cyber sphere and possible cooperative measures. Six GGEs have been convened since 2004, although some concluded without consensus and no final reports. Key outcomes from the 2014/2015 GGE report included the affirmation that States, in their use of ICTs, must observe State sovereignty, settle disputes peacefully, non-intervention in internal affairs and comply with international law, including human rights obligations. States must also not use proxies for internationally wrongful acts using ICTs and should prevent their territory from being used by non-State actors for such acts. All those concepts are retained as basic blocks in the current international discussions.

More recently, UN established a second parallel track for discussing ICT security: the "cyber" Open-Ended Working Group (OEWG). Open to all Member States delegations, the cyber-OEWG also implements a wider "multistakeholder" approach including the participation of representatives of private sector, academia and civil society. The current process OEWG 2021–2025 aims towards a future UN single-track mechanism to unify the double track process GGE-OEWG. The 3rd Annual Progress Report (APR) of the cyber-OEWG was adopted by consensus in July 2024, covering topics like existing ICT threats, responsible state behavior, development of Confidence Building Measures (CBMs) and capacity-building and a proposed Program of Action (PoA) for a UN permanent, global, inclusive coordination mechanism. An important initiative of the cyber-OEWG is the implementation of a Point of Contact (PoC) directory, a global, intergovernmental initiative collecting technical and diplomatic references to facilitate communication and coordination on cybersecurity issues and incidents.

International debate also includes the application of Jus in bello (International Humanitarian Law, IHL) and Jus ad bellum (limits on the resort to force, UN Charter Article 51) to the use of ICTs.

Regarding specifically AI, in July 2023 the UN Security Council held its first-ever meeting on threats to international peace and security, highlighting major risks, particularly concerning autonomous weapons and nuclear weapons control. The UN Secretary-General emphasized AI's potential for human development while cautioning against malicious use, raising political, legal, ethical and humanitarian issues.

In March 2024, the UN adopted the first resolution on Artificial Intelligence. In this context the Secretary-General warned against "sleepwalking into a dystopian future where the power of AI is controlled by a few people, or worse, by opaque algorithms beyond human understanding", emphasizing the need for rules, safety, and universal guardrails".

There is growing support for international action also regarding Lethal Autonomous Weapon Systems (LAWS), where AI plays a fundamental role. In 2017 an Open Letter addressed to the UN from various technology companies and academic representatives raised an alarm about LAWS becoming the "third revolution in warfare". These weapons could enable conflict at unprecedented scale and speed, be used against innocent populations and hacked. This letter also stressed the urgency for preventive actions. The UN Secretary-General has called for the conclusion, by 2026, of a legally binding agreement to prohibit lethal autonomous weapons systems, non-compliant with IHL.

8.8 Conclusion

The increasing integration of ICT, AI and unmanned systems is profoundly transforming warfare and challenging international security. While offering potential benefits like enhanced efficiency and safety, these technologies also present significant risks: the proliferation of autonomous weapons raising ethical and legal dilemmas, the potential for AI to destabilize nuclear deterrence, the inherent vulnerabilities of AI systems exploitable by adversaries and the pervasive use of information operations to manipulate populations.

The development of increasingly powerful AI systems introduces new levels of complexity and risk, including opacity, vulnerabilities and the potential for unintended or malicious applications like designing bio-chemical weapons. The ability to bypass ethical filters and the existence of unregulated models exacerbate concerns about control and trust. The blurring distinction between military and civilian domains and the potential for information manipulation to erode public trust, pose fundamental challenges to societal stability.

The international community, particularly through the UN, is actively engaged in addressing these issues, focusing on applicability of international law, developing norms for responsible state behavior in cyberspace and initiating discussions on AI's impact on peace and security. There is a wide recognition of the urgency to regulate autonomous weapon systems and establish global guardrails for AI. The challenges of ensuring Meaningful Human Control over autonomous systems and establishing clear accountability remain central.

The path forward requires continued international dialogue, transparency, confidence-building measures, and potentially legally binding instruments to manage the risks posed by these rapidly evolving[3] technologies.

The question of whether we can trust and truly control this technology underscores the critical need for timely and decisive action at the international level.

[3] This paper describes the status of AI technologies as of spring 2024, the current hyper-exponential evolution may change drastically the situation in a relatively short time, hopefully driving to a more positive landscape from a human, technical and international point of view.

Chapter 9
Domestic Nuclear Debate in Russia

Nadezhda Arbatova

Abstract This article analyzes the views and role of the main schools of strategic thought involved in Russia's domestic political debate on nuclear weapons. The main topics in this debate are revolving around three issues: "can a nuclear war be fought and won?", possible resumption of nuclear testing, and prospects of nuclear proliferation.

9.1 Introduction

Let me start with two preliminary remarks. First, under the circumstances Russian nuclear debate is not an issue of local proportions but rather of the Hamlet-like dilemma to be or not to be. It means that the issue of domestic nuclear debate goes far beyond Russian borders and it is central to international security.

Second, this topic is a logical continuation of my previous presentation. However, there are some fundamental differences in terms of the main factors that shape the course of the debate. The previous nuclear debate prior to Russia's special military operation in Ukraine was guided by the variety of considerations—marginalization of nuclear problems after the end of bipolarity, emergence of new disruptive technologies, crisis of arms control regime and etc.

Nowadays the main driving force of the domestic nuclear debate is the Ukraine conflict, which is characterized by the hybrid warfare—attrition war and exchange of Russian and Ukrainian missile attacks on their cities in line with the war of the cities during the Iraq-Iran war. This combination is particularly dangerous. As Winston Churchill said "The war of attrition will be ended by the exhaustion of nations rather than the victories of armies". As for the war of the cities, it cannot but lead to an escalation of the conflict, which may erupt into an all-out war.

N. Arbatova (✉)
Doctor, Professor of Political Sciences Head of the Department of European Political Sciences
Primakov National Research Institute of World Economy and International Relations (IMEMO), Russian Academy of Sciences, Moscow, Russia
e-mail: narbatova@mail.ru

P. Cotta-Ramusino et al. (eds.), *Nuclear Weapon Risks in a Problematic Time*, Springer Proceedings in Physics 358, https://doi.org/10.1007/978-3-032-29873-7_9

9.2 The Main Schools of Russian Strategic Thought

Domestic nuclear debate in Russia is going between representatives of the two main schools of strategic thought: one sometimes called— "nuclear fire worshippers" [9] and the so called school of strategic realists.

The first school is supposedly based on the Faculty of World Economy and International Affairs, at the Higher School of Economics in Moscow. Its ideological leader is a well-known political scientist, Professor Emeritus Sergei Karaganov, who has a number of followers and disciples like political analyst and Research Professor at the Higher School of Economics Dmitry Trenin, a group of military experts mostly in the field of geopolitics, some media figures like journalist Dmitry Suslov, co-host of the TV program "The Big Game" and others.

The school of "strategic realism" is considered to be based at the Centre of International Security at the Primakov Institute of World Economy and International Relations at Russian Academy of Sciences led by academician Alexey Arbatov. Many widely known strategic experts belong to this school –Major General (retired)Vladimir Dvorkin, a Chief researcher at the Centre of International Security; academician Sergey Rogov, former Director of the US and Canada Institute (who passed away in 2025); Major General (retired) Pavel Zoltarev, Deputy Director of the US and Canada Institute; Ivan Timofeev from Russian International Affairs Council (RIAC) and others.

In between there is a collection of intermediate positions of those experts who do not want to take sides or to openly state their position. Most of them are retired military like Lieutenant General Evgeny Buzhinsky and Colonel Michail Khodarienok. On the one hand, they argue that selective use of nuclear weapons can be justified, at the same time they recognize that escalation cannot be controlled and there could not be winners in a nuclear war.

9.3 Fundamentals of Russian Nuclear Debate

The nuclear debate in Russia raises many issues but it is mainly revolving around three points: (1) can nuclear war be won?, (2) resumption of nuclear testing, (3) prospects of nuclear proliferation.

"Nuclear war can be won and should be fought?"

If we remove the ideological husk from the statements of «the nuclear fire worshippers» about the Anglo-Saxon five hundred hegemony, "strategic parasitism of the West" betrayal of Christian values and the like, then we shall be left with a very clear message: "A nuclear war can be won and should be fought." [10].

They justify their nuclear posture by the constatation that the restraining fear of the atomic bomb, which existed throughout the second half of the twentieth century, is gone in the West. Nuclear deterrence no longer works. So, the practical conclusion is obvious for them: it is necessary to bring back the fear of a nuclear war and to prove that Russian nuclear deterrence which implies the use of nuclear weapons is not a bluff [11, 13].

This belief defines a broad framework for their views on when, how and where nuclear weapons can be used. Some of the representatives of this school insist that that Russian leadership should modernize its nuclear deterrence strategy, taking into account the practical experience of the Ukrainian conflict. This implies expanding the plans of nuclear weapons use" [11, 13].

Other protagonists of this school claim that Russia cannot win the war in Ukraine unless it demonstrates an evident willingness to use tactical nuclear weapons not only against Ukraine but selected European states. Namely they say that "this military operation cannot end with a decisive victory without forcing the West to retreat strategically, or even surrender, and compelling it to give up attempts to reverse history and preserve global dominance… We must go up the deterrence-escalation ladder quickly enough" [5]. They suggest that Russians and Belarusians living in Ukraine or elsewhere should be evacuated by Russia to prepare for launching a tactical nuclear weapon. The way to identify those nationals and the ways of their evacuation are not explained.

Proposed steps along the escalation ladder towards nuclear war are built on the assumption that the West will be unwilling to respond. "As for possible Russian nuclear strikes on NATO countries, hypothetically speaking, Washington is unlikely to respond to these strikes by attacking Russia for fear of its retaliation against the United States. The absence of such a reaction will dispel the myth built for decades around Article V of the North Atlantic Treaty and will lead to the deepest crisis in NATO, perhaps even to its collapse. It cannot be ruled out that the Atlantic elites in NATO and EU countries will panic and will be swept away by national forces, which will realize that the security of their countries does not depend on the non-existent U.S. "nuclear umbrella," but on building equitable relations with Russia" [11, 13]. They recognize that there will probably be some sort of response from Washington, but it is unlikely that the Americans would sacrifice Boston for Poznan, just as they were not going to sacrifice Chicago for Hamburg during the Cold War" [12].

They admit that Russia will have to expect severe condemnations not only from the West but also from the rest of the world, including China and India. However "in the end, the winners are not judged. And the saviors are thanked". It is the victor in the war who decides what is right or wrong. Russia has a divine duty to defend its place in history [5].

The nuclear posture of the "nuclear fire worshippers", first and foremost their reliance on nuclear prevention, is refuted by the school of "strategic realists" who put forward several counterarguments. First, they state that nuclear deterrence still works, otherwise NATO's long-range missiles would be delivered to Ukraine and NATO troops might be already fighting in Ukraine. Still, nuclear deterrence is not a panacea or a guarantee against major wars because it can be undercut by the escalation of the conflict.

Second, they say that the appeal to use nuclear weapons violates the nuclear taboo, which has been in place and respected since Hiroshima and Nagasaki. Once that taboo has been broken, controlling further escalation would be impossible. The appeal to "activate the nuclear factor in world politics" would not prevent, but on the contrary, inevitably increase the likelihood of a nuclear war.

Third, the assumption that Russian nuclear first use can take place in Ukraine or elsewhere in Europe without provoking further nuclear attacks is speculative. After crossing this Rubicon, it will become much more difficult to stop. All stages of a possible escalation are unpredictable.

Using tactical nuclear weapons on the part of Russia would result in an overwhelming attack using conventional weapons against Russian assets, including the Baltic and Black Sea fleets [3]. Russia would then have to respond with more numerous nuclear strikes against NATO. In response to this attack, NATO would almost certainly resort to nuclear retaliation against targets in Russia. So, selective escalation would in fact lead to a nuclear disaster.

As Alexey Arbatov has noted, "Tactical nuclear weapons are not some "sterile instrument." The bombs that wiped out Hiroshima and Nagasaki would now be classifies tactical grade. Further escalation through the stages of air-space strikes and nuclear attacks would quickly bring the conflict to an exchange of massive strikes with all the monstrous consequences [2].

The political damage for Russia would be enormous. For non-nuclear states, NPT members, Russia would immediately become a toxic actor with whom no one could afford to interact. Breaking the nuclear taboo could destabilize nuclear deterrence in other regions: i.e. between India and Pakistan, China and India, China and Taiwan, Israel and Iran. And this would be the final convincing argument for all those states that are striving to go nuclear.

9.4 Resumption of Nuclear Testing

As the war in Ukraine has ground increasingly towards stalemate, Russian foreign policy experts and media figures have pushed for a new kind of escalation: resumption of nuclear tests. It is no coincidence that the first touchstone regarding the resumption of nuclear tests was thrown by the representatives of the school of "nuclear fire worshippers."

Aside from the exotic proposal of a journalist Margarita Simonyan to test a nuclear weapon "some place over Siberia" and send a "nuclear ultimatum" to the West, [7] more sophisticated political experts from this school of thought supported the idea of "resuming nuclear testing as soon as possible: first underground, and if this is not enough, then with the detonation of Tsar-Bomba-2 on Novaya Zemlya, while taking steps to minimize damage to the environment of our own country and of friendly World Majority states [6].

Interestingly, these ideas have been supported by some well-known Russian physicists but in line with president Putin's position who "admitted that Moscow would conduct nuclear tests only if Washington does it". Russia and the U.S. continue to abide by the Comprehensive Test Ban Treaty (CTBT), which established an international legal norm against nuclear testing. But the treaty has not come into force: nine countries have not ratified the CTBT including the US and Russia. The latter withdrew its ratification in November 2023. It means that the CTBT has not incorporated previous treaties on nuclear testing. For whatever reason, this fact is being ignored by the proponents of nuclear test resumption.

Russian strategic realists emphasize several important logical connections between these treaties. First, the CTBT is inextricably linked to the NPT. Calls for a prohibition on nuclear testing have been a central tenet in the nuclear field since before the Treaty on the Non-Proliferation of Nuclear Weapons (NPT). The aspiration to ban all nuclear testing is reflected in the preamble of the NPT and in the Article VI.[1] In 1995, part of the package of decisions that ensured the indefinite extension of the NPT was the commitment to conclude CTBT negotiations by 1996. Early entry into force of the CTBT was one of the 13 Practical Steps, agreed upon by NPT States Parties in 2000 [8].

Second, Each State Party has the right to withdraw from the CTBT with six months' notice, which means that Russia's withdrawal would not surprise anybody. With its withdrawal from CTBT Russia could conduct only underground tests because the Limited Test Ban Treaty signed by the United States, the Soviet Union, and Great Britain in 1963, banned all nuclear tests in the atmosphere, in space, or underwater. Underground nuclear tests can be registered by seismic detectors, which won't have the effect that is the purpose of this act.

If Russia complies with the procedure of withdrawal from the mentioned international treaties, the United States would also be ready to resume nuclear testing. As in the past, an unprecedented arms race will begin; in which Russia will have a very difficult time. The "nuclear fire worshippers" ignore history when they refer to the testing of Tsar-bomba in 1961, but not to the international consequences of this test, first and foremost, an intensifying nuclear arms race and international tensions that led to the Cuban missile crisis of 1962 and eventual Khruschev's missiles' withdrawal.

9.5 Prospects of Nuclear Proliferation

Nuclear fire worshippers" sometimes openly approve further nuclear proliferation. They argue that sooner or later, Russia will have to change its official position on this issue, which was imposed by Americans who wanted to minimize their risks. Still non-proliferation policy was unfair to many non-Western states and stopped working long ago.

Moreover, they claim that "historically and philosophically, proliferation contributes to peace". It is frightening to even imagine what would have happened

[1] "Each of the Parties to the Treaty undertakes to pursue negotiations in good faith on effective measures relating to cessation of the nuclear arms race at an early date and to nuclear disarmament, and on a treaty on general and complete disarmament under strict and effective international control".

if the USSR and then China had not developed nuclear weapons". Having acquired nuclear weapons, Israel became more confident among its hostile neighbors. But if Israel's neighbors had nuclear weapons, Israel would have acted more modestly. Having carried out nuclear tests, India has become more secure in relations with a more powerful China. North Korea is more confident and is raising its international status, especially since Russia has finally stopped dragging itself in the wake of the USA and de facto resumed cooperation with Pyongyang" [6]. They say that limited nuclear proliferation may also prove useful as a barrier to the creation and use of other weapons of mass destruction. But most importantly, nuclear weapons, including their proliferation, are necessary to restore the aspects of nuclear deterrence that have ceased functioning—to prevent not only major conventional wars (as in Ukraine). A conventional war cannot be won if the potential enemy has nuclear weapons and, most importantly, is ready to use them.

Summing up, according to this school of thought, "the polycentric and stable world order of the future will not come without nuclear multilateralism…". However, a "selective" approach is being put forward here: it is necessary to help Iran, the United Arab Emirates, Saudi Arabia, Egypt, as well as Brazil and Mexico with the acquisition of nuclear weapons. But if Germany or Poland try to destroy these weapons, they should be destroyed "with a preemptive strike, and the same applies to Japan. Naturally, proliferation also carries risks. But given the current disorder and redivision of the world, these risks are much smaller than those that result from the weakening of nuclear deterrence [6]."

Contrary to the prevailing view that nuclear weapons are a reliable guarantee against big wars, the "strategic realists" say that nuclear proliferation is a recipe for disaster [1]. With horizontal nuclear proliferation the likelihood of nuclear conflict increases as the number of nuclear states with completely different nuclear strategies expands. Moreover, Karagnov's proposal that Russia should help its allies to build nuclear weapons is a double-edged sword. The same logic could be applied by the West to Ukraine or Taiwan. The Ukraine President Volodymyr Zelensky during his October visit to the US met former President Donald Trump and told him that Kyiv will pursue nuclear weapons for defense if it doesn't gain entry to NATO [4]. Prior to the Budapest Memorandum, Ukraine had the world's third-largest nuclear stockpile on its soil following the breakup of the Soviet Union. It is one of four countries—along with Belarus, Kazakhstan and South Africa—to have voluntarily embraced nuclear disarmament. It seems that the appearance of a nuclear power on Russia's borders is unlikely to meet Moscow's interests. How would the worshippers feel?

9.6 Conclusion

One of the most important questions is whether there is a feedback link between the nuclear debate of experts and the official course of the Russian leadership. And if there is, what is the impact? Of course, there is no "iron curtain" between Russia's experts' community and the leadership. Unlike the Soviet time, the current nuclear debate has a range of differing views. On the one hand it is impossible now to find in this debate those ideas about a nuclear free world or nuclear disarmament, which were very popular 50 years ago. On the other hand, few experts call to launch a massive nuclear strike against the United States. These are the limits.

Representatives of the Russian political elite attend the most important forums like the Primakov Readings and the Valdai Forum. Recently President Putin directly debated the nuclear issues with Karaganov. However, like in other countries the causal relationship here is not obvious. It is not crystal clear who is exerting influence on whom. Probably, in our case it may have the opposite vector: creating a benign propaganda background for a more moderate course adopted at the top. The published amendments to the nuclear doctrine do not indicate a fundamental revision, but rather a clarification of some of its provisions. Contrary to the proposals of the radicals, officially Kremlin is not talking about preventive or pre-emptive nuclear strikes, limited nuclear war, or the use of nuclear weapons as a utilitarian means of achieving the success of military operations in Ukraine and beyond.

Despite wide spread views that all red lines are blurred in the Ukrainian conflict, these lines still exist. For the West, this is the first use of tactical nuclear weapons by Russia. For Russia, it is the West's permission for Ukraine to hit Russian cities, primarily Moscow and St. Petersburg, with long-range weapons as well as the possible acquisition of nukes by Ukraine. In this regard, speculations that exist in both Russia and the West that the parties are bluffing when they talk about the possibility of a nuclear response to each other's actions are very dangerous. We often hear from the Western experts and the Russian opposition located abroad that Kremlin will never use nuclear weapons in the Ukraine conflict and such statements are just blackmail. The same with regard to the US and NATO we hear from some of our fire worshippers. Such perceptions intentionally or unintentionally push the sides towards a nuclear escalation.

A threat assessment results from a focused and exhaustive research effort and is expressed in probabilities instead of certainties. So, the issue is not whether "I believe down the street, the probability that a brick will fall on your head is very small. But if you are walking on a construction site, the probability will be much higher, and you should wear a safety helmet. A nuclear threat assessment is thus a specialized area of expertise that requires understanding of a wide range of issues, including escalation lines and even the place of analogies in political and military thinking. The answer to the question of whether president Putin wants to use nuclear weapons will be "no", he does not want to. But could there be a situation in which he would use nuclear weapons (like any other leader of a nuclear power)? Yes, there could be such a situation. Hence it is obligatory to prevent such cases.

References

1. A. Arbatov, We need to talk not about disarmament but about restoring arms control (Interfax, 2024). https://www.interfax.ru/russia/964173

2. A. Arbatov, Nuclear metamorphoses (August 8, 2023). https://russiancouncil.ru/en/analytics-and-comments/analytics/nuclear-metamorphoses
3. A. Arbatov, K. Bogdanov, D. Stevanovich, Nuclear war is a bad way to solve problems (Kommersant, 2023). https://www.kommersant.ru/doc/6055340
4. D. Glebova, Zelensky says he told Trump Ukraine will pursue nuclear weapons if it can't join NATO (New York Post, 17 Oct 2024). https://nypost.com/2024/10/17/us-news/zelensky-says-he-told-trump-ukraine-will-pursue-nuclear-weapons-if-it-cant-join-nato/
5. S.A. Karaganov, A difficult but necessary decision, Russia (Global Affairs, 13.06.2023). https://eng.globalaffairs.ru/articles/a-difficult-but-necessary-decision/
6. S.A. Karaganov, An age of wars? Article two. What is to be done. Russia. Global Aff. **22**(2), 5–8 (2024). https://doi.org/10.31278/1810-6374-2024-22-2-5-8
7. D. Kaverin, The Kremlin estimated Simonyan's idea of a nuclear explosion in Siberia (Gazeta.ru, 2023). https://www.gazeta.ru/politics/news/2023/10/03/21418315.shtml
8. G. Mukhatzhanova, 75 years after trinity: the nexus between the CTBT and the NPT (VCDNP, 2020). https://vcdnp.org/75-years-after-trinity-the-nexus-between-the-ctbt-and-the-npt/
9. I. Rinaeva, You are right to be afraid of a nuclear war (Komsomoloskaja Pravda, 2024). https://www.kp.ru/daily/27634.5/4984521/
10. M. Rostovsky, Nuclear war can be won, in Moskovsky Komsomolets, October 11, 2023. https://karaganov.ru/en/nuclear-war-can-be-won/
11. D. Trenin, Conflict in Ukraine and nuclear weapons, Russia (Global Affairs, 2023). https://eng.globalaffairs.ru/articles/ukraine-and-nuclear-weapons/
12. D. Trenin, The U.S. and its allies are playing 'Russian Roulette;' you'd almost think they want a nuclear war (RIAC, 2022). https://russiancouncil.ru/en/analytics-and-comments/comments/the-us-and-its-allies-are-playing-russian-roulette-you-d-almost-think-they-want-a-nuclear-war/
13. D. Trenin, Bring nuclear weapons back into the strategic equation (Strategic alert Service, 2023). https://www.strategicalert.news/en/dmitry-trenin-bring-nuclear-weapons-back-into-the-strategic-equation/

Chapter 10
Israel Between Iran's Regional Ambitions and Nuclear Plan

David Menashri

Abstract The Islamic Revolution was a turning point in the history of modern Iran, marking a change in the regime's ideology and a shift in its policy. In its attitude toward Israel, Iran moved from close friendship to bitter animosity. In the past, the countries were distant from each other, but Iran's increased involvement in the region, directly or through its proxies, turned it into a kind of neighbor with which conflicts intensified. Iran's nuclear program has added critical dimensions to the enmity. Since April 2024, the conflict has escalated from a shadow war to mutual attacks, creating a significant shift in the balance of power in favor of Israel. This article examines Iran's policy toward Israel, focusing on two crucial areas—regional policy and the nuclear program. Finally, it examines their bilateral relations following the Hamas attack on October 7, 2023, until the end of 2024.

10.1 From Close Friendship to Bitter Animosity

The legacy of Iran is based on three main pillars—ancient, imperial, monarchical, and Persian; medieval, Islamic-Shi'ite; and the third, which has become entrenched over the past two centuries: new, modern, Western. The three layers joined together to shape the new Iran. Vacillating between the imperial heritage, Islamic tradition, and an anti-imperialist approach proved to be a complex experience, a mixture of Iranian nationalism, Islam, and the West. The first two are anchored in Iran's sense of uniqueness and superiority over its neighbors and the aspiration to be a major power in the region [8, 5]. The influence of each layer is shaped by the changing realities and the unique perceptions of the leadership at any given time. Iranian nationalism is often typified by disdain and arrogance toward its neighbors, and by the belief that Iran and Shi'is have been unjustly deprived of their deserved central role in world history. The senses of victimization and "historical humiliation" strengthen the aversion to foreign interference, which was reinforced by the perceived American hostility [1].

D. Menashri (✉)
Alliance Center for Iranian Studies, Tel Aviv University, Tel Aviv, Israel
e-mail: menashri@tauex.tau.ac.il

P. Cotta-Ramusino et al. (eds.), *Nuclear Weapon Risks in a Problematic Time*, Springer Proceedings in Physics 358, https://doi.org/10.1007/978-3-032-29873-7_10

Caught between the challenge of effectively running the state and ideological convictions, the Islamic regime felt often obliged to adopt a pragmatic policy. In 1987, Ayatollah Ruhollah Khomeini sanctioned the authority of the state even "to destroy a mosque" or suspend the exercise of the "five pillars of the faith," if the state's interest so required (Kayhan, 7 January 1988). Like other ideological movements in history, upon assuming power and facing the complex demands of governance, the new regime was often forced to adapt itself to newly emerging realities. How does Iran's assertion that there is no difference between Muslims, either in ethnic or sectarian affiliation, accord with the constitution, which states that only a Shi'i of Iranian origin can hold office as president? How can one reconcile the abhorrence of national divisions within the world of Islam with the insistence that the Gulf must be called Persian? The revolutionary ideology negated the existence of Israel, but unlike in most other cases, the regime did not see yet a sufficient interest to deviate from its dogma.

Iran and Israel do not share common borders, have had no wars between them, and have no territorial claims on each other. Although Iran recognized Israel only de facto (in 1950), they evolved into an intimate strategic alliance—"relations of love without a marriage contract," as a senior official described it to me in 1976. This "Golden Era" came abruptly to an end in 1979. The revolutionary ideology and its strategic ambitions posed severe challenges to Israel. Iran's nuclear program, along with its advanced long-range missiles, were additional concerns. On its part, Israeli animosity toward Iran, its operations against the nuclear program, and its regional policy made it the archenemy in Iranian eyes.

Iran's Islamic arguments against Israel placed the hostility on a religious rather than a political-national basis. According to its dogma, Judaism is not a nation but a religion; Jews do not deserve a state, certainly not in Palestine. The conflict was seen as one between righteousness and falsehood, and it is the "religious duty" of Muslims to fight it. Israel was viewed as the power behind every sinister scheme. Moreover, since Islam is the last monotheistic religion (and due to the abundance of oil in their lands), Muslims are merit to be *the* leading force in human society. Furthermore, instead of Jews being a tolerable minority in Muslim lands, many Muslims now live under Jewish rule. Thus, history has gone off the rails, making it the duty of the Islamic Republic to correct this distortion. Israel's close relations with the Shah were also held against it. Israel was thus the enemy of Iran, of Islam, and of humanity, and therefore should be "wiped out from the map." More practically, for Shi'i Iran to claim leadership of the Muslim world, it must hoist the Palestinian flag and lead the struggle against Israel. Milder expressions were infrequent and usually voiced by people with no official standing. Some argued that, regardless of revolutionary slogans, Israel could not be eliminated, that the Prophet made agreements with the Jews, and advised that Iran, as a Persian maxim maintains, should not be "a bowl that is warmer than the soup"—more Palestinian than the Palestinians. For his part, Supreme Leader Ali Khamenei viewed Israel as "racist," a "cancerous tumor," created to foil Muslims' unity, honor, and integrity. The "final solution" Khamenei proposed is that all Palestinian refugees should return to Palestine and together with Jews who lived there before 1948 hold a referendum to approve an [Islamic] state democratically

(Radio Tehran, 20 October 2000). In Persian, this approach is called "strangling with cotton"—killing softly.

During the first decade of the revolution, regional policy was defined principally in ideological terms. Gradually (mainly since 2009), ideology was placed in the service of Iran's national interests and security. The new realities encouraged a more balanced foreign policy, based on rational-pragmatic assessment [2]. In both areas, domestic and regional developments have been insufficient to produce a breakthrough. In the first years after the revolution, when Iran was engaged in war with Iraq, Iran even bought arms from Israel, in 1985/6 (the Iran-Contra affair).

Iranians then rationalized this attitude by claiming that to avoid drowning, it is admissible to hold on to the snake's tail. As in Iran, Israel's tone became much more heated at the beginning of the millennium. Israel took a firm line against Iran on every issue: human rights, animosity toward Israel and Jews, Iran's nuclear project, and its expansion in the region. Israel volunteered to head the front against Iran's nuclear program and made a worldwide effort to promote awareness of Iran's threat to Israel, the region, and the world. It also worked to thwart the nuclear program and acted against nuclear sites and scientists. This has significantly slowed down Iran's plans, but did not stop the nuclear project. Although after the nuclear agreement (2015), Iran was faithful to its obligations, it advanced its plans after the United States' unilateral withdrawal from the agreement (2018) and during the negotiations with the United States on its renewal in the early 2020s.

By the end of 2010, it had become evident that Israeli targeted killings of nuclear scientists, sanctions, and cyber-attacks had failed altogether to stop Iran's progress. Their plans for an air attack on Iran didn't materialize until 2024 (except for a few targeted attacks). According to former Mossad Chief Meir Dagan, Prime Minister Benjamin Netanyahu was the only Israeli prime minister that "the entire defense establishment failed to accept his position" regarding a military attack [3]. Another former Mossad Chief Tamir Pardo blamed the high level of uranium enrichment Iran had reached on a lack of strategy on the part of Israel, along with the Trump administration, which had led Iran to be become recently potentially only weeks away from a nuclear weapon (Jerusalem Post, 9 June 2021). Israel's failure to block Iran's regional expansion resulted from numerous factors: underestimating Iran's ability to build support among proxies; a mistaken belief that its enemies were deterred; and overconfidence that it could overcome any threat due to its military might (mainly air superiority, rockets and advanced technology), combined with a typical arrogance (see below).

The advantages of Iran's use of proxies are clear: the dissemination of religious-political doctrine; recruiting faithful yet inexpensive military personnel; supporting friendly movements in sensitive locations; and gaining influence and advancing its interests beyond its borders. Proxies also provide a convenient zone of deniability; their use is cheaper and saves lives on the patron's side. The chain of proxies also enabled a land corridor from Iran to the Mediterranean and a pathway to the strategic Bab al-Mandeb Strait.

The use of proxies, however, is not without problems. Not all proxies fully adhere to Iran's policy, or share its ideology and goals, and their commitment to Iran is contingent on their own interests being met. Sharp disagreements with Syria, their main ally, arose already during the Iran-Iraq War, and in the 1990s Syria entered peace negotiations with Israel, despite Iran's huge investment in Syria. For its part, Iran never promised to always support its proxies. The use of proxies also led to increased cooperation among Iran's opponents (mainly within the Sunni axis and strengthening relations with Israel).

The US and European countries also opposed the proxies and imposed sanctions. Moreover, the countries from which militia fighters were recruited (such as Afghanistan and Pakistan) were displeased by the recruitment of their citizens and their deployment as "cannon fodder" in service of Iran's interests. At home, the astounding expenses spent on proxies were severely criticized. Also, the need to recruit foreign combatants attested to a diminished willingness for sacrifice among Iranian youth. Iran has also spent billions of dollars annually supporting Syria and its proxies [4, 7].

Iran developed a strategy of "Forward Defense" based on proxies around its border (Lebanon, Syria, Gaza, and the West Bank) and further away (Iraq, Yemen, and Bahrain), as well as on strategic ports in Yemen and a land route to the Mediterranean Sea. Contact between them was maintained through the Iranian-led "resistance front." The movements are diverse in terms of religious-sectarian affiliation (mainly Shi'is from different schools of thought as well as Sunnis), ideology, interests, Iran's role in their formation, and their ties with Iran.

Hezbollah is the "flagship" of the creed to export the revolution. It is often dubbed the "jewel in the crown" due to its ideology and close ties with Iran, its location on Israel's border, and its military might - an asset for the moment of truth. It is loyal to Iran's revolutionary doctrine and recognizes Khamenei as the source of religious authority. However, it has been gradually integrated into the Lebanese governmental system and cannot fully detach from the state's interests.

Thus, it has proven sometimes reluctant to confront Israel to spare Lebanon from its counterattacks. In Iraq, following the call of Ayatollah Ali al-Sistani, Popular Mobilization Forces were established in 2014, and various militias joined its ranks. Some of their leaders also hold key positions in the government and thus must consider state interests. There are also independent Shi'ite organizations in Iraq, such as the popular organization of Muqtada al-Sadr, headed by the Saraya al-Salam militia, which opposes Iran's policy and supports an extremist anti-American line. In Yemen, the main force is Ansar Allah (Houthis), a Zaydi Shi 'i movement that differs from other proxies in terms of ideology, formation, and commitment to Iran. They seized control of the country in 2014 and control Sanaa and large parts of Yemen, including strategic port cities. The Houthis' alliance with Iran was forged when, in the early 2010s, the Iranians recognized their potential to serve Iran's interests against rivals in the Gulf region and Israel. Iran assists them with armaments, guidance, and funding, but they are not its pawns and do not agree with it on every issue. For their part, they are resentful of what they see as Iran's lax policy vis-à-vis Israel (see below).

Iran managed to save the Assad regime against all odds until December 2024, tightened its relations with China and Russia, cooperated with Turkey, despite previously tense relations, and increased its influence in the Horn of Africa. Hezbollah has become the main force in Lebanon, while the militias operating under Iranian aegis are the dominant force in Iraq. Hojjat al-Islam Mehdi Ta'eb, close to the Supreme Leader, even went so far as to describe Syria as Iran's 35th province. For him, Syria was more important to Iran than Khuzestan. He rationalized that if Iran retains Syria, it will be able to retake Khuzestan; but if Syria falls, Iran will lose Tehran. This view bluntly ignored the heavy sacrifices of this province in the war, and the fact that Iran has only 31 provinces. Only by counting Iraq, Lebanon, Yemen, or Bahrain could the total reach 35 provinces ('Asr-e Iran 16 February 2023).

While the regime was building on its "successes," public fury has increased, especially since the late 2010s. Many have opposed Iran's adventures beyond the borders, the sacrifices in lives and money in support for allies and proxies. A popular maxim, frequently heard in Iran, linked popular distress to support for proxies: "A lamp essential for the household is forbidden [haram] to donate it to the mosque!" Others set against religious and revolutionary maxims practiced in Iran, preaching instead devotion to Iran: "neither Islam, nor Qur'an, I will sacrifice my life for Iran."

Thus, while the slogan of the 1979 Revolution was "Islam is the solution," an attractive and catchy phrase with which many could identify, by 2009 many wondered, "Is Islam the solution?" More recently, many resolved that Islam, as practiced under the Islami rule, is not the solution.

Criticism also came from senior officials. Already in 2015, Rouhani stated that, "the main enemy" of Iran is not the U.S., nor Israel, but unemployment, inflation, drought, and crisis of ecology (Radio Free Europe\Radio Liberty, 18 September 2015). In a video clip on the eve of the Persian New Year in 2020, former Foreign Minister Mohammad Javad Zarif condemned the gaps within Iranian society. He courageously advised "conceptual remodeling" and "regime innovation" (Ferraro, 2020). Former President Mahmoud Ahmadinejad criticized the regime for concentrating all powers but shifting responsibility to the people. He concluded bluntly: This is not religion of Allah, but religion of Satan (Doulatebahar.news, 16 November 2021; YouTube, Zolmnews, 17 November 2021). It was long clear that the main problem in Iran is at home.

10.2 Iran and the War in Gaza

The Hamas attack on 7 October 2023, followed by the ongoing war in Gaza and clashes with the "resistance front" and Iran itself, has had far-reaching effects on the Middle East and beyond. By the end of 2024, it is still not clear when and how the war will end and what its implications will be for the various players. Yet, indisputably, it has changed the regional architecture. Iran's initial policy remained faithful to its regional strategy: support for Hamas and the Palestinian Islamic Jihad (PIJ), while denying any involvement or prior knowledge of the attack. It aimed to avoid responsibility, minimize damage, and safeguard its interests. Concurrently, it aspired to avoid confrontation with the U.S. and to avert normalization between Israel and the Sunni axis, particularly with Saudi Arabia.

The proxies were an important tool to advance its goals, not the other way around. Iran glorified Hamas' "heroic victory," denied prior knowledge of the war (its timing, not the plan), and offered its help as "part of the solution"—positioning itself as both arsonist and firefighter. Iran applauded Israel's military and moral defeat, which in its view shattered the myth of Israeli power and invincibility, while emphasizing that the attack was carried out exclusively by Palestinians. Yet, senior officials approved that Iran provided Hamas with funding, arms, and training, and had significant part in its plans. According to Kayhan, the day before his assassination (3 January 2020), Gen. Qasem Soleimani presented the "resistance front" with a treaty for cooperation (Kayhan, 2 January 2020). Iran also formed an "operations room" for the Front to coordinate between Tehran and its proxies (Tabnak, 20 October 2023).

In the early stages of the war, the proxies declared their support for Hamas and backed it with attacks on Israel. First and foremost was Hezbollah, which harshly criticized Israel but sought to avoid a major war that would involve the U.S. and damage Lebanon's national interests. Hezbollah had also to consider, the interests of Lebanon and the discontent among parts of the public over the damage caused by its reckless actions. All of this necessitated a degree of restraint. In the meantime, Israeli communities near the border were severely damaged, both sides suffered significant losses, and tens of thousands of Israelis from its northern border became refugees in their own country. Many residents of southern Lebanon were also forced to flee their homes. The attacks on both sides of the border escalated until, by the end of September 2024, the conflict had entered a new phase. The Israel Defense Forces (IDF) launched a series of assassinations in Beirut, including that of Nasrallah and several Hezbollah leaders, alongside Israeli attacks and IDF ground maneuvers in southern Lebanon.

On 11 October 2023, Ayatollah Sistani expressed support for the Palestinian struggle but stopped short of calling for armed resistance against Israel. Iraqi Proxies' leaders held key positions in Mohammed Shia al-Sudani's government, which necessitated restraint. Sadr focused on presenting maximalist demands (mainly against the U.S and Israel), such as a call to end the U.S. military presence in Iraq and the Middle East more broadly. Washington's initial response to missile attacks on American installations from Iraq was limited to strikes in Syria rather than Iraq and was measured to avoid destabilizing Iraq or prompting the withdrawal of U.S. forces stationed there to counter ISIS. However, the United States later moved to attack Iraq. Prime Minister al-Sudani warned the U.S. against attacking Iran, fearing that political pressure might force him to take undesirable decisions. On 28 January 2024, pro-Iranian militias in Iraq attacked an American base in northeast Jordan, killing three U.S. military personnel. Both Iran and the U.S. made it clear that they had no intention of entering direct confrontation, and Khamenei even instructed the Iraqi militias not to link Iran to the activity in Jordan.

The Houthis, however, were far more eager to participate in the war due to their radical anti-Israeli ideology and their ambition to be seen as a major regional force.

Threats to navigation in the Red Sea, mainly against ships en route to Israel, prompted several major international shipping companies to reroute their vessels, significantly impacting global supply chains and costs. In response, the U.S. established an international task force to ensure safe passage through the vital shipping routes. Yemen, often dubbed the "backyard" of the world, has become a significant ideological-military force and a key player against Israel and the U.S. The Houthis operate with increasing independence, reflecting a partnership rather than a proxy relationship. They are not "puppets" controlled by Iran via "remote control," but partners and allies. Even if Iran wanted to restrain them, it is uncertain whether it could (Washington Post, 16 May 2016; Foreign Affairs, 29 January; Iran Foreign Ministry Website, 25 January 2024).

Tensions have also emerged with Syria. Assad was the only leader of the "resistance front" who did not congratulate Iran for its attack on Israel on 13 April, did not participate in Jerusalem Day ceremonies, and was absent from Raisi's funeral. Assad, who received billions from Iran and owes it his survival, chose instead to grant economic benefits to Russia, to Iran's disappointment. Strategic gaps have also emerged between Iran and its allies, as diverging interests began to threaten the coherence of its proxy strategy. Assad pursued a relatively independent policy; Iraqi militias attacked American bases; and the Houthis adopted an aggressive posture toward Israel and the United States, carrying out strikes on U.S. warships and on Israel in response to Israeli strikes on Sanaa and Hudaydah Port.

The Houthis did not hesitate to harshly condemn Tehran for not responding to Israeli attacks. A member of the Houthi Council, complained that while the proxies shed their blood, those who don't want to expand the scope of the campaign [i.e., Iran] refrain from action. He advised bluntly: "either do something useful or shut up!" (*TRT World*, 26 December; Cratersky.net, 27 December 2023). As a foreign observer noted, Iran seems "like a chariot drawn by a band of unruly horses." Iran holds the reins, but its allies often decide the pace and direction (Wall Street Journal, 19 August 2024).

A meeting in Tehran between IRGC commanders and the proxies turned into a heated confrontation. The Guards called for restraint until the outcome of the ceasefire talks in Gaza became clear and insisted that only Iran could make the final decision. The proxies, however, pushed for extensive action, including strikes against the U.S. While Pezeshkian and his associates advocated calm, some IRGC generals demanded attacks on military bases in Tel Aviv (The Telegraph and Ynet, 9 August; Al-Jaridah, 14 August 2024). To balance, Pezeshkian said that Israel is an apartheid regime, commits genocide and war crimes, ethnic cleansing, possession of nuclear weapons (Tehran Times, 12 July 2024).

The Iranian attack on Israel from its territory marked a transition from a "shadow war" to direct confrontation. An Israeli strike on an Iranian diplomatic facility in Damascus, regarded by Tehran as Iranian territory, on 1 April triggered pressure from hard-liners and some proxies for a strong response. Khamenei, who usually favored restraint over adventurism, faced a severe dilemma: a forceful reaction risked escalation that could destabilize his regime, while inaction could provoke a credibility crisis at home and among allies. Iran responded with a massive aerial assault on Israel

(13–14 April) and promptly announced that this was its revenge for the Damascus attack, and that it considered the matter as closed (Ynet, 14 April 2024). The attack, though spectacular, was an embarrassing failure. On 19 April, Israel retaliated by striking Iranian military sites, including targets near Isfahan, including nuclear facilities. Israel also severely damaged the air defense systems in Syria, Iraq, and Iran. Iran did not retaliate swiftly, due to its military inferiority. Gen. Amir Ali Hajizadeh (commander of the IRGC's air and space force) stated that the soldiers were eager to strike, but "Iran's hands are tied" (SNN.ir, Khabargozari Daneshju, 1 July 2024).

Lethal Israeli attacks on Beirut in September 2024—some resembling scenes from science fiction—inflicted severe damage on Hezbollah, created confusion in Iran and among its proxies, and restored some of Israel's damaged prestige and self-confidence. On 17–18 September, thousands of pagers and hundreds of walkie-talkies were exploded, killing or wounding numerous Hezbollah operatives. On 29 September, an Israeli strike on a bunker where Nasrallah was meeting with senior officials killed all attendees. On 1 October, Iran again attacked Israel from its territory, though even less ostentatiously than in April. On 16 October 2024, Yahya Sinwar (head of Hamas' political bureau) was killed by the IDF in Rafah. On the night of 25–26 October, Israel struck numerous air defense targets in Iran, Iraq, and Syria, that together with the attack in April creating a reality of "open skies" for future attacks. These actions gave Israel a qualitative edge in the war against the proxies, especially against the Iranian nuclear program.

Israel seemed eager to strike Iran's nuclear infrastructure but still refrained mainly due to its prolonged multi-front war, hostages still held in Gaza, and the need for U.S. collaboration, or at least green line. Meanwhile, as the Israel-Iran duel played out, Iranian officials suggested that any Israeli strike on its nuclear sites would necessitate a review of Iran's nuclear doctrine. Intelligence assessments indicated that Iran's nuclear program was then better positioned to produce nuclear weapons, should it decide to do so.

The election of Donald Trump in 2024 gave Israel a meaningful tailwind. Saudi Arabia, Egypt, and other Arab states declined to adopt Iran's stance toward Israel. When Khamenei claimed that the war in Gaza had brought global attention to the Palestinian cause, Mahmoud Abbas, head of the Palestinian National Authority, countered that Iran's "goal is to sacrifice Palestinian blood … The Palestinians do not need wars that do not serve their ambitions for freedom and independence" (Khamenei.IR, 24 March; Palestinian News Agency, 3 June 2024). In Lebanon, a ceasefire was reached between Israel and Hezbollah. Clashes between Israel and the proxies in Iraq and Yemen significantly subsided. In December, the Assad regime collapsed, and the land bridge that Iran had once boasted of became "a bridge to nowhere." The fundamental problems remain unresolved.

Between the completion of this article and its publication in summer 2025, momentous developments have taken place. The "12-Day War" (13–24 July 2025) between Israel and Iran had far-reaching effects on the region and beyond. Israel emerged triumphant and regained some of its prestige and self-confidence. The Iranian missile strikes on military bases and civilian targets (including scientific institution and hospital) caused extensive damage but did not change the general balance of power. Israel took advantage of its aerial superiority, technology and intelligence to acquire complete control of Iran's skies, lead massive attacks on strategic targets and nuclear facilities, and assassinate senior nuclear scientists, top commanders and attack regime symbols. On the 12th day, the United States attacked the key parts of Iran's enrichment infrastructure in Fordow, Natanz and Isfahan, using B-2 Spirit stealth bombers (carrying bunker-busting bombs) and Tomahawk cruise missiles at Natanz and Isfahan to complement the bombing campaign. Israel was euphoric while Iran—in a panic. The United States, having achieved its military goals and to avoid further spread of violence, rushed to reach a ceasefire that Iran was eager to achieve unconditionally, and Israel accepted reluctantly.

Despite all the impressive military success, there are still unknowns: Were the nuclear sites obliterated (as President Trump insists) or only severely damaged (as Iran maintains)? How much of the enriched uranium was demolished or moved to a hiding place? While undoubtedly the attacks delayed the nuclear plan, it is not clear what the future bears: would the sides reach a new agreement? would Iran abandon its pursuit of nuclear plans? Or would it attempt to accelerate—after some time—the nuclearization of Iran?

As far as Iran is concerned, it has suffered a severe blow and faced difficult challenges—increased economic distress, popular disenchantment, decline of its regional standing and the loss of popular support. The regime tried to present a façade of "business as usual" and attempts to emerge from the initial shock, but it does not look the same as it did the day before. Under all accounts, the regime is challenged at a most vulnerable moment (both at home and away).

One thing seems to be clear: In the pendulum between the centrality of religious against national identity, at this point the latter seems triumphant—a painful reality for the Islamic Revolution. The young have long distanced themselves from the initial revolutionary ideals and turned more to Persian nationalism. Not only that, the regime, including Khamenei, has also begun to cling to the Iranian national past.

As far as Israel is concerned, despite significant success, the war in Gaza is not over, Israelis are still being held hostage in Gaza, and the promised "absolute victory'" has yet to be achieved. The rifts in society have not been healed, and the political rivalries have deepened. The damage of the October 7th attack and the war since then will tack long to remedy.

10.3 Conclusion

In the two main foci of this paper, expansion of Iran's influence in the region and its nuclear program, by October 7th Iran has achieved significant success. Yet its status as the omnipotent leader of the Resistance Front has declined since the war and was shattered in April 2024. Israel, for its part, has not succeeded in depriving Iran of its assets in the region but has not succeeded in stopping its nuclear plan by 2024. Despite heroic operations by its military-security apparatus and the impressive resilience of its civil society since then, the Israeli desired "absolute victory" has been achieved. It will take long to repair the damage and return evacuees to their homes, not to mention the heavy/casualties it suffered. Over fifty abductees are still held in Gaza, alive or dead—their return remains a top priority for most Israelis. The Palestinians, it should be recalled paid even heavier price.

Israel's security doctrine since its foundation has rested on three main fundamentals: early warning, deterrence, and decisive victory. Yet in 2023 it was caught in total surprise and suffered both a military and cognitive blow, and the myth of its invincibility was undermined. The voice of Netanyahu, who had long led the campaign against Iran's nuclear threat, has been muted since his return to office in early 2023, when he became absorbed in the judicial reform. It remains unclear whether or how Israel will be able to stop Iran's nuclear ambitions. The jubilant declarations about restored deterrence following tactical successes, and the celebratory tone in television studios, would have been better replaced with humility and modesty. They have shown how little did Israel learn from the Hamas attack.

It should be recalled that Iran's growing influence in the region did not stem from the appeal of its revolutionary ideology or from securing freedom and welfare for its own people. Iran's main regional successes occurred where ruling systems had collapsed, become dysfunctional, and left a power vacuum such as in Lebanon, Iraq, Syria, Yemen, or Gaza. Iran has proven well adept at fishing in murky waters. Yet, it built its proxy network to serve Iranian interests, not to waste their war potential in such attacks. Such a move would contradict Iran's strategy and risk U.S. intervention and a regional war, with heavy cost. Iran has invested heavily in its proxies over decades and did not intend to squander them to serve others' interests. The Resistance Front was created to act in the service of Iran's interests on the day of the order, and not so that Iran would get involved in a campaign that was not its own and at an inopportune time. Activating the resistance front is not a "maneuver" that could be repeat again and again. Its failure threatened to crumble the front.

The absence of a strategic vision was a major flaw in Israel's regional policy, its approach to Iran's nuclear program, and its handling of the war in Gaza. Tamir Pardo observed that Israel has "top-notch tactics, but zero strategy" (Jerusalem Post, 9 June 2021). Reuel Gerecht and Ray Takeyh added that, for Israel, "accumulating tactical victories is a strategy." One important gain available to Israel is the chance to connect with a bloc of more moderate Sunni states, mediated by the United States. It is not certain this would have succeeded, but Israel did not give it the attention it deserved. In a single day, Sunni and Shi'i Islamists unraveled an emerging Sunni-Israeli alliance (The Free Press, 2 April 2024).

It is often said that "it takes two to tango." In the case of the Israeli Palestrinian conflict, there is not even one side that seems ready to make the necessary compromise for this "dance." Israelis often scoff that the Palestinians never miss an opportunity to miss an opportunity. Today, there are those who say the same about the Israelis themselves.

In the West, when something is dismissed as irrelevant, people say, "The rest is history." But in the Middle East, everything is history. The increasingly inescapable lesson of the Islamic Republic's relations with Israel, wrote Prof. Ali Ansari, is that there is too much politics in the history, and not enough history in the politics.

Until this imbalance is addressed, the chances for meaningful progress remain slim (Ansari, A., Foreign Affairs, 29 May 2024).

References

1. M. Ahouie, Developments of Iranian political discourse on Israel. From past to present. Discourse **9**(3–4), 57–70 (2011)
2. K. Barzegar, The geopolitical factor in Iran's foreign policy (Middle East Institute, 2009)
3. R. Bergman, Rise and kill first: The secret history of Israel's targeted assassinations (Politico Magazine, 2018)
4. H. Da'i, Gozaresh vidio'i az hazineh mali-ye hemayat as rezhim-e Assad [Report on the budget for protecting Asad's regime] (Khabarnameh Guya, 2017)
5. F. Farhi, Crafting a national identity amidst contentious politics in contemporary Iran, in *Iran in the 21st Century: Politics, Economics and Conflict*, ed. by H. Katuzian, H. Shahidi (Routledge, London, 2008), pp. 13–27
6. R. Khomeini, Al-Hukumah al-Islamiyah [The Islamic Governance] (1970)
7. A. Lane, Iran's Islamist proxies in the Middle East. (Woodrow Wilson Center, Washington DC, 2023)
8. M. Sariolghalam, Iran in search of itself, in *Current History* (2008), pp. 425–431

Chapter 11
Perspectives on a Two-State Solution After the Recent Escalation in the Middle East

Taghreed El-Khodary

Abstract As Any reasonable person would find it absurd to talk about a two-state solution when a genocide is taking place—when Israel burned families alive, when Israel continues to use starvation as a weapon, when Israel continues its policy of wiping out entire families, when Israel declaring its intend of ethnic cleansing. So far, over 54,418 Palestinians have been killed, among them more than 17,000 children. In July 2024 the world-renowned medical journal *The Lancet* published that 8% of Gaza's population has perished—killed by bombs, starvation, lack of medical care, and disease. That was July 2024. The rate is higher by now. Israel banned UNRWA—the United Nations Relief and Works Agency for Palestine Refugees—from operating in the West Bank, East Jerusalem, and Gaza. Came out with "an alternative" Gaza Humanitarian Foundation, turns to be Gaza Humiliation Foundation. Sadly the US is a partner in this humiliation tool that aims to turn The Gaza Strip into unliveable place. To this date, Israel refuses to accept a ceasefire, despite the overwhelming international calls. Let's remind ourselves here of what were the Political solutions options that we used to discuss as journalists, analysts before the plausible genocide we are witnessing with our eyes live from Gaza.

11.1 The Two States Solution

For over 30 years, the Oslo Accords have been the cornerstone of this proposed solution. But Oslo failed. The key issues—borders, East Jerusalem, and the right of return for Palestinian refugees—were postponed indefinitely. These core issues were not discussed from day one. Postponement led to the failure. Let alone, Israel keeps from the time of Oslo years till today expanding its illegal settlements, undermining any hope for a viable Palestinian state.

T. El-Khodary (✉)
Palestinian Commentator and Media Expert, Brussels, Belgium
e-mail: taghreed.elkhodary@gmail.com

P. Cotta-Ramusino et al. (eds.), *Nuclear Weapon Risks in a Problematic Time*, Springer Proceedings in Physics 358, https://doi.org/10.1007/978-3-032-29873-7_11

Another option we talked about was *An Egalitarian Binational State*. This would be based on mutual legitimacy and equal rights between two national groups—Palestinians and Israelis. But this option was unacceptable to both Israel and the West. Israeli practices on the ground at that time obviously intended to fail Oslo. This Egalitarian Binational State option gained more support among Palestinian intellectuals and those opposing Oslo and the youth.

If Israel is not interested in the two-state solution or Binational State, are they interested in *Apartheid State*? Its policies made it clear to us that they are ok with the apartheid state. Israel then moved to establish the *Abraham Accords*. Instead of pursuing a just resolution between Israelis and Palestinians, Israel embraced normalization with few Arab states through the *Abraham Accords*—completely excluding Palestinians. The Palestinian cause was pushed out of the political discussion, and the Palestinian Authority (PA) was reduced to a security subcontractor, with no control over its own people's fate. PA to remain as the biggest employer responsible for only one thing: Security cooperation with the Israeli military occupation. The PA to guarantee security for the Israeli occupation by arresting Palestinian resistance groups. The outcome post the Abraham accords is zero recognition of Palestinian self-determination. Palestinian cause is completely pushed out of the political discussion. Meanwhile, Gaza under siege and what tuned to be wishful thinking is turning Hamas into a municipality.

Gaza has been under siege since 2007—17 years of collective punishment. After winning the legislative elections, Hamas was denied political legitimacy, and Gaza was placed under complete siege.

Then came October 7. A few members of Hamas sought to change the status quo, interrupting a possible Abraham accord with the Saudi Arabia. Also, an attempt to releasing the thousands of Palestinian political prisoners in Israeli jails. Hamas attack was shocking to all. Gazans' immediate reaction was fear of Israel's response. They have lived through four wars and the Second Intifada. But no one anticipated what followed: an unprecedented genocide, mass displacement, and a continuation of the 1948 Nakba.

11.2 Voices from Gaza

A major Every day for over a year,
my family and friends in Gaza ask me:

Is there any hope that Israel will stop?
Why has the world abandoned us?
Will the ceasefire talks succeed this time?

All of my relatives have been forcibly displaced, except for Zuhair, my brother-in-law, who still lives in his home in Gaza. One of the oldest houses in Gaza city. He refuses to leave. "Where would I go?" he asks. "Look what happened to those who left. I prefer to die with dignity in my home."

My dearest cousin Hind remains, reporting for Al Jazeera English from the middle areas of Gaza. She refuses to flee, committed to her ethical responsibility.

Other distant relatives and neighbours, the Khurshid family, also refused to leave. The mother insisted that everyone stay in their home in Gaza City. Imagine this: all who stayed inside—her son, her grandchildren—were killed by an Israeli airstrike. Only she, the grandmother, survived, simply because she happened to be on the balcony when the missile hit. The injured grandmother managed to leave Gaza to the US to be with her two daughters. Two of her grandchildren survived because they were out of the house during the Israeli bombing.

We have witnessed humiliation, killing, displacement, and dehumanization—and the world remains "concerned." That's all we hear.

How can anyone speak of a two-state solution while Israel continues its settler-colonial project in Gaza, the West Bank, and East Jerusalem? President Biden, who calls himself a Zionist, has refused to use the significant leverage the U.S. has over Israel. Instead, the only advice given is: "*Try to minimize Palestinian deaths.*" And now same American policy continues with President Trump.

11.3 Western Hypocrisy and Media Lies

There are two lies that drive me mad when repeated by mainstream media:

1. "*Israel has the right to defend itself.*"

Israel is defending only one thing: its military occupation.

2. "*Hamas uses Palestinians as human shields.*"

In reality, Israel forces Palestinians into so-called safe zones—then bombs them. Israel is using Palestinians as human shields and the media does not cover it as it is.

I am a Palestinian from Gaza, born and raised in the Rimal neighborhood, near the now-destroyed Shifa Hospital. I went to schools that are now rubble. I worked in media offices that have been bombed. I drank coffee at the Dirah café by the Mediterranean—it no longer exists. Hamas never used my family, neighbors, or community as human shields. They never fired rockets from our areas. Israel's objective is clear: ethnic cleansing and the extermination of Palestinians. Freedom of expression is not antisemitic, it is essential. We must never fear speaking the truth. The Palestinian cause is just that is where we draw our strength.

11.4 What Must Be Done

End the occupation

The root of the problem isn't Hamas or Hezbollah—it is the military occupation. We need a political solution based on international law and UN resolutions. Israel says no to everything—no to peace, no to boycott, no to resistance. So what does it want?

Enforce the ICJ ruling

On July 19, 2024, the International Court of Justice ruled that Israel's occupation of Gaza, the West Bank, and East Jerusalem is unlawful and must end. Settlements are war crimes and must be dismantled. Western powers must not be complicit and repeat history.

11.5 Demand Accountability

Israel must be held accountable for its crimes. Without justice, the cycle continues. As ICC prosecutor Karim Khan recently revealed, he was told the court was "built for Africa and for thugs like Putin," not for Western allies. This mindset must be rejected.

End Western hypocrisy

Western governments cannot selectively apply human rights. Some even praised a deadly Israeli attack in Lebanon. How have they reached the point of celebrating civilian deaths/terrorism?

Impose an arms embargo

The U.S. and Europe must stop sending weapons that fuel this genocide.

Fix media bias and treat Israeli officails same way you treat Russian officials

Journalists must provide context and resist Israeli propaganda. If Israel continues to deny access to Gaza for International journalists, then media outlets should consider boycotting interviews with Israeli officials, who only promote their propaganda narrative. A form of pressure is never too late. It's never too late to push back with facts and confront the Israeli politicians with their crimes. International Journalists must be allowed to go to Gaza and document all.

11.6 Conclusion

Both the West and the Arabs have failed to stop the genocide in Gaza, to stop the pogroms in the West Bank and the expansion of the illegal Israeli settlements in the West Bank. The question I keep asking why could the West and the UN impose a solution on Serbia, thus creating Kosovo. WHY NOT IN PALESTINE? How come allowing extremist/fascist government in Israel to carry out these crimes. How come the West is failing again after promising never again.

Chapter 12
The International Quest for Fusion, Past and Future

Francesco Pegoraro

Abstract Controlled thermonuclear fusion may provide the long term solution to the process of energy transition and prove to be an energy source that benefits the whole of human society. Research on its realization in the laboratory started several decades ago but is still far from commercially viable energy production, even if very promising results have been obtained in recent years. The quest for net energy production in the laboratory has been hindered by physics and technology challenges, but these very difficulties have provided the stimulus for a strong international collaborative effort that has crossed ideological and political barriers and has provided a major channel for peaceful agreements and nuclear arms control through the direct contact between scientists from the different sides.

12.1 Introduction

The aim of laboratory research into thermonuclear fusion is to be able to reduce the process that keeps stars alight to the scale of a power plant [1]. The energy that stars radiate originates from fusion reactions in which, differently from nuclear fission, the nuclei of two light elements fuse to form a heavier nucleus, releasing energy in the form of kinetic energy of the particles produced. To achieve the fusion of the two nuclei it is necessary to provide them with enough energy to overcome the electrical repulsion that would keep them apart. The physical conditions for this process to lead to a net energy gain are those of a high-temperature confined plasma, of the order of a hundred million degrees, where fusion reactions can compensate for all energy losses [2].

There are two main types of confinement: magnetic confinement and inertial confinement. In magnetic confinement the plasma is enclosed inside a toroidally shaped magnetic "bottle" and heated until it is brought to the required conditions.

F. Pegoraro (✉)
Physics Department, Pisa University, Pisa, Italy
e-mail: francesco.pegoraro@unipi.it

Accademia Nazionale dei Lincei, Rome, Italy

P. Cotta-Ramusino et al. (eds.), *Nuclear Weapon Risks in a Problematic Time*, Springer Proceedings in Physics 358, https://doi.org/10.1007/978-3-032-29873-7_12

In inertial fusion a small pellet of solid matter is compressed and heated by very powerful laser pulses. In the fusion reaction that has the highest reactivity and that thus will be used first to produce energy, a Deuterium and a Tritium nucleus fuse forming a Helium nucleus (an α-particle) with the additional release of a neutron. For equal reacting masses this process produces about ten million times more energy than in a chemical combustion process. In both magnetic and inertial schemes the energy produced will be extracted with a heat exchange process from a Lithium blanket. Like Deuterium, Lithium is easily found in Nature and when hit by neutrons from the fusion reactions produces the Tritium needed to compensate for the burnt fuel.

12.2 Recent Developments

There have been important new developments in recent years with promising results achieved both in inertial and in magnetic fusion. In December 2022 at the US National Ignition Facility (NIF) for the first time more fusion energy (3.15 MJ) was produced than was supplied to the Deuterium–Tritium (D–T) pellet (2.05 MJ), thus starting a burning process [3]. This was a very significant milestone even if the energy supplied to the pellet was only a fraction of the energy needed to produce the laser pulses. In the framework of magnetic fusion an important result was obtained in the UK at the Joint European Tokamak (JET), one of the longest-running and most successful experiments in the field of fusion. In December 2021, in a plasma discharge lasting over 5 s, 59 MJ of fusion energy were produced[1] corresponding to a fusion power of ~ 11 MW. The energy produced exceeded, even if with a lower "gain" (significantly below energy breakeven), the previous record set at JET in 1997.

In addition to these energy records, there have been significant advances in the science and technology of magnetic fusion in countries outside Europe and the US, in particular in China, Japan, India and Korea. In China the EAST tokamak produced a "quasi-steady" plasma pulse with a record duration of 1056 s. In Japan JT-60SA, an advanced joint Japan-Europe experiment, was inaugurated in December 2023.

From the physics and technology point of view a very important recent development is the renewed interest in the "high magnetic field line" (up to ~ 20 T), rekindled by the recent realization of large-size "high temperature" (~ 20 K) superconducting magnets.

However these developments do not guarantee a significant shortening of the time required for the realization of a *commercial* nuclear fusion plant. Indeed, in the case of the magnetic fusion project International Thermonuclear Experimental Reactor (ITER),[2] we are perhaps witnessing a further delay.[3]

---

[1] https://ccfe.ukaea.uk/fusion-energy-record-demonstrates-powerplant-future/.

[2] https://www.iter.org.

[3] https://www.world-nuclear-news.org/Articles/ITER-director-general-promises-realistic-project.

12.3 Nuclear Fusion for Peace

Producing fusion energy in a controlled way and at laboratory scales has presented practical and conceptual physics challenges that were initially not anticipated. Interestingly, the very difficulties in achieving net energy production have been the main stimulus, at least in the case of magnetic fusion, for a *strong, and now world-wide, international collaboration.* This collaboration has lasted for more than half a century through the cold war and, as far as we can see, the recent tensions due to the war in Ukraine. In the case of inertial fusion such a wide collaboration has been partly hindered because of its indirect connection with nuclear weapon stewardship. Such a connection is not present in the case of magnetic fusion that instead has developed a strong relationship with astrophysics and with space physics in particular.

The idea that it might be possible to obtain net energy in the laboratory from controlled nuclear fusion processes predates WWII, but the effective birth of fusion research dates from the early '50s [4]. It was part of the general endeavor of scientists in the nations that had achieved nuclear weapon capabilities to use atomic energy (thus including fission and fusion) for peaceful energy production. In their research fusion scientists experimented with a range of different magnetic configurations trying to address first the problem of confining the plasma in a stable way and then of heating it. These efforts were carried out in strict secrecy with researchers in different nations ending up following unknowingly similar experimental approaches and thus finally encountering similar difficulties. Direct contacts between scientists on the two sides of the iron curtain made this evident and eventually led to the conclusion that international collaboration was necessary in order to speed up the development of fusion: in 1956, the Soviet physicist Igor Kurchatov gave a talk at the Atomic Energy Research Establishment at Harwell in the UK where he described the Soviet fusion program and detailed the problems they were facing.[4] Since the Soviet scientists that the imposed classification was initially meant to keep in the dark were at roughly the same stage of development there was no real reason to maintain it.

The international collaboration on fusion research obtained a major endorsement at the Second Conference on the Peaceful Uses of Atomic Energy in Geneva in 1958[5]:

> This problem [thermonuclear fusion] seems to have been created especially for the purpose of developing close cooperation between the scientists and engineers of various countries". (L.A. Artsimovich, from the Introduction to the Conference and, following J.D. Cockcroft[i] one of the vice-presidents of the Conference), "The discussion was greatly helped by the final stage of declassification of this field of research by the US and the UK. The U.S.S.R. presented four volumes containing one hundred papers describing the results of their fusion research up to the present time, and also gave some information about their recently completed large-scale fusion devices". [5]

---

[4] https://discovery.nationalarchives.gov.uk/details/r/C2038429?descriptiontype=Full&ref=PREM+11/1611.

[5] https://digitallibrary.un.org/record/3808553?v=pdf.

The IAEA *Fusion Energy Conferences*[6] became a major channel where technical and scientific information on progress in fusion was shared at an international level. However the conference in Culham in 1965 made it evident that progress along different experimental lines had stalled while, still with the memory of the fiasco[7] of the ZETA experiment in the UK, results reported by the Soviet scientists that indicated a significant improvement in plasma heating in toroidal devices were met by skepticism.

A major step forward was made at the end of the '60s when in view of additional positive results (plasma temperatures in the "tokamak" T-3, an order of magnitude higher than in other devices were reported at the 1968 IAEA Conference) the Soviet scientist Lev Artsimovitch invited a team from the UK to the Kurchatov Institute in Moscow in order to verify their temperature measurements independently. The positive scientific (and human) outcome of this scientific mission that validated their temperature measurements [6] represented a game changer in fusion research and led to the subsequent adoption almost worldwide starting from the '70s and continuing in the '80s of the tokamak concept for plasma confinement and heating.

It was soon realized that part of the difficulties encountered due to the plasma reduced energy and density confinement in the laboratory were caused by the onset of plasma instabilities that gave rise to "anomalous" transport coefficients and required a kinetic (as opposed to fluid) description in regimes that had not been explored before.

In the '60s a basic understanding of the nature, onset and expected effects of these instabilities was established through a wide international collaboration based on open (translated) literature and direct personal contacts. The International Center for Theoretical Physics (ICTP) created in 1964 in Trieste played an invaluable role. It had a strategical location, Soviet scientists were allowed to stay in the West remaining close to the (then) Yugoslav border. It was a golden period for plasma physics and fusion science. Sagdeev and Galeev's Trieste "Lectures on nonlinear plasma theory" guided the development of the field for the next 20 years. Quoting from F. Kennel's personal account [7] "*The chance to hear Rosenbluth and Sagdeev lead a debate on the state of plasma physics attracted to Trieste a talented group of researchers, who all were or became prominent*". In the following years "*Rosenbluth and Sagdeev used their friendship to advance international arms control, and Sagdeev became one of Soviet Premier Gorbachev's principal arms control advisors*", a proof of how important direct contacts between fusion scientists were in opening new channels for advancing peace.

The '80s saw two large tokamak experiments, aimed at reaching energy breakeven: (the condition when the power released by the fusion reactions equals that required to heat the plasma). The Tokamak Fusion Test Reactor (TFTR) at Princeton Plasma Physics Laboratory (PPPL) funded at a national government level, and the Joint European Torus (JET), funded by an Inter-governmental agreement within the framework of the EURATOM organization.

---

[6] https://nucleus.iaea.org/sites/fusionportal/Pages/fec-landing.aspx.

[7] https://sciencenotes.org/today-science-history-january-24-project-zeta/.

Neither of these experiments reached breakeven but both obtained significant fusion energy production in D–T experiments and led to major advances in energy confinement and plasma heating, in collaboration with scientists and engineers from around the world. Soviet research continued on the tokamak line but started to feel the consequences of growing economic difficulties leading in the '90s to the diaspora of its scientific community. In these same years more nations and institutions in Europe, in the Americas and in Asia (Japan, India and later South Korea and China) started to join active fusion research, or strengthened existing activities, bringing, in particular within the Asian context, their different political visions. Again most enmities, such as the one between India and China, were at the governmental level and did not stop contacts and collaborations between scientists that acted as future political bridges. Fusion research gradually stopped being dominated by its three initial main players, USA, UK and USSR and started to grow as a full world-wide enterprise.

The Strategic Defense Initiative (SDI) was announced by President Ronald Reagan in 1983. Because of the possible resulting loss of a balancing power SDI was a cause of concern for the USSR. US President Ronald Reagan and Soviet General Secretary Mikhail Gorbachev met in Geneva in November 1985 at the height of the cold war tension in order to examine the possibility of reducing the number of nuclear weapons. However "preparations for the summit showed that there were no readily available agreements on weapon reduction, while the project of a renewed joint effort in fusion research was strongly supported on both sides by Alvin Trivelpiece (Director of the Office of Energy Research in the '80s) and by Evgeny Velikhov (future head of the Kurchatov Institute). The two scientists supported a project to construct a demonstration fusion reactor and believed that taking the next step in fusion research would be beyond the budget of any of the key nations and that collaboration would be useful internationally".[8] The last item of the final Geneva joint statement *"emphasized the potential importance of the work aimed at utilizing controlled thermonuclear fusion for peaceful purposes and, in this connection, advocated the widest practicable development of international cooperation in obtaining this source of energy, which is essentially inexhaustible, for the benefit of all mankind."* At the summit in Reykjavik in 1986, the two leaders reiterated their commitment and officially launched the international fusion initiative. Five months later US, Soviet, European and Japanese representatives named the project ITER.

12.4 A Large Scale International Enterprise

ITER construction was approved in 2006[9] after lengthy discussions (started in 1992 under the auspices of IAEA and concluded in November 2007) on its design and on its location. Presently it is under construction: its envisaged start of Research

[8] https://en.wikipedia.org/wiki/iter.

[9] https://www.iter.org.

Operation is in 2034 while its first D–T operation is not expected earlier than 2039. ITER does not aim to reach ignition, i.e. the physical condition when the fusion power is sufficient to keep the plasma burning without any external energy source. The aim is a factor 10 in fusion power above breakeven.

The group of ITER partners has expanded: China and South Korea joined in 2003 and India in 2005. The position of the US has changed (now it is again a partner), that of the UK (after Brexit) is at the moment unclear.

Independently of any judgment on the validity, from the point of view of physics and technology, of the approach adopted or of the risk of concentrating most of the resources on a single, highly complex experiment, ITER may be seen as providing an illustration of different aspects of the international collaboration in fusion research.

A definite positive aspect is the cultural and scientific bridges it contributed to establish between politically divided countries. China and Japan collaborate as partners within ITER and the collaboration has extended to other scientific fields. There are bilateral formal collaboration frameworks on fusion research between Japan and China and between Japan and Korea. Indian and Chinese scientists work together as partners within the ITER project even if there is no collaboration program between China and India. Nevertheless, Chinese and Indian scientists work together in international forums such as the International Tokamak Physics Activity.

A negative aspect is the complexity of the ITER organization that has its roots in the way in which the project was initially formulated so as to reach agreement through what could be called "scientific diplomacy".

This has led to delays in the deliveries, incompatibilities and thus to further delays in the development of the project in addition to those that could be expected from its sheer technical complexity.

At the same time international collaboration between politically rival countries remained active in these very recent difficult years: "Project Center ITER" (Rosatom's enterprise) has carried out one of the largest shipments of equipment for the fusion mega-facility in August 2024.[10] It is to be hoped that such a collaboration may continue in future years notwithstanding last years geopolitical developments.

12.5 Present Changes

The periodic revisions towards longer times of the ITER schedule and the realization that it might be safer to have a wider approach to the search for a possible future magnetic fusion reactor (including approaches different from a tokamak such as the stellarator line) pointed towards the need to proceed at a faster pace.

New projects were started aimed in a number of cases at addressing specific scientific or technical issues. In most cases these projects were conceived at a national

[10] https://rosatom.ru/en/press-centre/news/rosatom-ships-this-year-S-largest-batch-of-equipment-to-the-iter-construction-site/?sphrase_id=6399625.

level, but international collaboration continued to play a major role e.g. through an active exchange of scientists.

A recent development in fusion research has been the entry of private initiatives (or at least partially financed by private capital) in the form of startup companies. This development has led to a multiplication of initiatives both in the field of inertial and magnetic fusion and reflects a process, already evident in other fields of technological innovation (e.g. communication satellites), where the role of governmental or international structures is taken in part by private ones.

Some of these startups follow approaches that do not appear to be well founded scientifically and almost all indicate times for the realization of a fusion reactor prototype that appear far too close. In some cases, however, the proposed programs are scientifically valid and innovative and these initiatives certainly have an important role in support and stimulation (see among many others Sparc in the USA[11] and Tokamak Energy in the UK.[12]

These private projects are not designed to address simultaneously and solve all the physics and technology problems that the development of fusion as an energy source requires. However, they can represent an opportunity to show in a shorter time that an energy source based on nuclear fusion has actually been achieved and, possibly, to indicate alternative approaches in the event that the current one based on large tokamaks proves not to be optimal for the construction of a fusion reactor.

The division between private and public funding is not always sharp, see the Italian DTT[13] consortium, or the inertial fusion startup Focused Energy[14] in Germany that has received large government funds. At the same time in Asia and in the West governments are strengthening national programs: the UK[15] and Germany[16] in particular plan to step up their investments in fusion research considerably.

A remarkable development is that in the last ten years China has taken a dominant role in fusion research in terms of investments, results and fast schedule, to the point of risking making existing experimental programs obsolete. However, in these years relations between China and the US have become strained. US researchers receiving federal funding have cut off official collaboration channels as requested by the US authorities, while a sort of collaboration exists between China and Taiwan that did not have a national fusion program. This collaboration is not official but it is in the open. Collaboration between China and European institutions has not been interrupted and in the Italian context we may recall the Center for Nonlinear Plasma Science (CNPS),[17] a "Virtual Center" based on a network of hubs, established to

[11] https://www.psfc.mit.edu.

[12] https://tokamakenergy.com/.

[13] https://www.dtt-project.it/index.php/organization/operative-structure.html.

[14] https://www.focused-energy.world.

[15] https://step.ukaea.uk.

[16] https://www.fusionindustryassociation.org/german-government-announces-new-fusion-funding-program/.

[17] https://www.afs.enea.it/zonca/cnps/index.html.

promote a joint research and collaboration framework that includes Italian, Korean and Chinese institutions and universities.

12.6 Commercial Fusion?

All these national and private activities bear witness to the belief that, although commercial fusion reactors (not just experiments in igniting plasmas or prototypes of fusion reactors) will not be able to supply power to the electric grid say for at least 40–50 years, nuclear fusion may represent the long-term solution to the problem of energy transition. With this in mind separate countries, as opposed to super-national institutions, and possibly also large companies already aim to capitalize on this future source of energy and in the meantime on its returns in terms of technological developments and applications (such as advanced materials, high temperature super-conductors). The positive recent experimental results mentioned at the beginning of this article have made the final goal "within sight" and the need for full international collaboration less imperative.

This change will come at a cost. Although fusion fuel, essentially Deuterium and Lithium, is relatively cheap and (probably also in the case of Lithium) geographically widely available, the development and the construction of a nuclear fusion reactor will have huge investment costs that will likely be affordable only by rich countries. Even more importantly, it will require a technical know-how that again will be available only in countries with a strong scientific and technological tradition or in countries capable of buying it.

There are two consequences: an obvious one is that also in the case of nuclear fusion, scientific education will be fundamental for the well being of a nation, a different one is that it seems unlikely that in the present social and political framework fusion energy will be a source of energy available equitably and affordably to all humankind.

12.7 Conclusion

Thermonuclear fusion on Earth is the process that empowers the destructive force of the Hydrogen bomb but, at the same time, it may contribute to solve the growing need of energy that is one of the main causes of conflict among human kind. Historically, fusion research has consistently promoted international collaboration and has been an instrument of peace. It is imperative to act in such a way that this remains true once net energy production by controlled thermonuclear fusion is finally realized.

References

1. National Academies of Sciences, Engineering and Medicine, in *Plasma Science: Enabling Technology, Sustainability, Security and Exploration* (Washington DC, 2021)
2. International Atomic Energy Agency, *Fusion Physics* (IAEA, 2012)
3. J. Tollefson et al., Nuclear-fusion lab achieves "ignition": what does it mean? Nature **612**, 597 (2022)
4. M. Barbarino, A brief history of nuclear fusion. Nat. Phys. **16**, 890 (2020)
5. J.D. Cockcroft, Peaceful uses of atomic energy. Nature **182**, 903 (1958)
6. N.J. Peacock et al., Measurement of the electron temperature by Thomson scattering in Tokamak T3. Nature **224**, 488 (1969)
7. C.F. Kennel, Rosenbluth and Sagdeev in Trieste; The birth of modern space plasma physics. J. Geophys. Res. Space Phys. **125**(4) (2020)

Chapter 13
The New Generation of Nuclear Fission Reactors

Bruno Panella

Abstract After the presentation of the energy context, showing the main issues related to primary energy production and the impact on the environment by the main energy sources, particularly concerning the CO_2 emission, the nuclear energy history is presented and, after some mentions to nuclear bombs, its pacific use is discussed. The fuel cycle and the new generation of nuclear power plants are described, with particular emphasis on the small modular reactors. The fuel cycle problems are also presented.

13.1 Introduction

Global primary energy consumption grew in the decade 2011–2021 from 12,441 to 14,208 million tons of oil equivalent (Mtoe), showing a growth of around 14%. Fossil fuel consumption remained largely unchanged at around 82% of primary energy consumption, that is the present percentage, but the produced energy, especially the electric energy, is higher. In the decade 2011–2021, global electricity production increased from 22,269 to 28,466 TWh, an increase of 28%, twice that of primary energy consumption. The following sources contributed to electricity production: 36% coal, 24% natural gas, 4% oil, 16% hydroelectric, 10% solar and wind, 10% nuclear power. The growth in energy consumption was driven by natural gas, which contributed more than 40% of the increase. China, the United States and India together contributed more than two-thirds of the global increase in natural gas consumption, with U.S. consumption expanding at its fastest rate in 30 years.

Almost half of the natural gas comes from the United States, due mainly to the so-called "shale" by means the "fracking," i.e. the hydraulic fracturing of the earth's crust with high-pressure jets that release the hydrocarbons imprisoned in the rock, allowing the United States to mostly achieve the energy independence. Russia, Iran and Australia are the other Countries that have contributed most to the growth of

B. Panella (✉)
Department of Energy, Politecnico of Turin, Turin, Italy
e-mail: bruno.panella@polito.it

P. Cotta-Ramusino et al. (eds.), *Nuclear Weapon Risks in a Problematic Time*, Springer Proceedings in Physics 358, https://doi.org/10.1007/978-3-032-29873-7_13

natural gas production. Concerning the electricity production per capita, USA occupied the first place with 12,235 kWh, followed by France, Netherlands, Germany, France, China, Italy and India. The European average is 5345 kWh. Global CO_2 emissions in the last two years amounted to more than 33 billion tons. The following Countries contributed most to the CO_2 emissions: China 10,523 billion, USA 4701 million, India 2552 million, Russia 1581 million, Germany 629 million. The concentration of CO_2 in the atmosphere has exceeded 400 ppm for some years, above the threshold of 350 ppm, which, according to a significant part of the scientific community, would avoid the worst effects of global warming.

The increasing electrification of the world economy has been accompanied by a growing contribution from renewable sources, which is one of the key actions in reducing carbon emissions. The share of renewable energy in electricity generation increased from 17.7% to 24.3%. Almost half of the increase in electricity is due to renewable energy, but in 2023 wind and solar power covered only about 3% of primary demand. Fundamental to limit CO_2 emissions is the increase in renewable energies (solar, wind, hydroelectric), but for the baseload, necessary due to their variability, if the combustion of fossil fuels is avoided, nuclear energy seems to be a choice to be considered, together with new innovative technologies, such as the storage, capture and reuse of CO_2.

13.2 Nuclear Bombs

The use of nuclear energy from the fission of Uranium and Plutonium started for military purposes during the Second World War (Manhattan Project). From September 1941 to the end of 1942 Fermi carried out experiments to obtain a controlled chain nuclear reaction. In 1942 the Manhattan Project started at Los Alamos (New Mexico), under the scientific direction of the physicist Oppenheimer. After the nuclear bombs on the Japanese cities of Hiroshima (developing a power of 12,500 TNT) and Nagasaki (developing a power of 20,000 TNT) in August 1945 (the USA President justified the use of the weapons with the fact that it had shortened the war and saved hundreds of thousands of human lives that would have been sacrificed if the conflict had continued), other Countries developed their nuclear arsenal, and over 2000 tests were conducted during the Cold War.

International Non-Proliferation Treaties (STARTs) have sought to limit the number of weapons. Russia and the United States hold 90% of the world's nuclear warheads. The new START treaty (which expires in 2026) limits both sides to 1550 warheads. Due to the war against Ukraine, Russia recently suspended it.

13.3 Fuel

Uranium is the fuel for nuclear power plants. In the Uranium mines there are two isotopes: U-238 (more than 99%) and U-235 (about 0.7%); U-235 is fissile for neutrons of whatever energy, while U-238 fissions for high energy neutrons. The fission of 1 kg of fissile material, in which all the nuclei are fissioned, produces the same energy as the combustion of approximately 1900 tons of oil. The proven and estimated Uranium reserves, at a price of $130/kg, amount to approximately 5.4 million tons. Recently the Uranium oxide production was about 60,000 tons, which corresponds to 76% of the world fuel demand, the remaining 24% was provided by reprocessed spent fuel, dismantled nuclear warheads, and extracted uranium reserves. The Uranium enrichment process consists of increasing the mass percentage of Uranium 235 to have sufficient fissile nuclei in the reactor.

This is a complex process because, since a chemical process cannot be used, a mechanical process is needed (e.g. ultracentrifugation). In most of the nuclear reactors the enrichment of Uranium 235 varies from 3 to 5%, but in the breeder reactors it reaches 15–20%. All reactors work essentially in the same way: Uranium or Plutonium splits into two fission products, emitting fast neutrons; some of these neutrons hit nearby atoms, causing them to fission; since these neutrons are fast, if there is only natural Uranium a chain reaction cannot be achieved only with natural Uranium. If a moderating substance such as water (natural or heavy) or graphite slows down the neutrons, they have more time to interact with other fissile atoms, so more atoms fission. If the fissile mass is sufficient, there will be a self-sustaining chain reaction. Nuclear energy is released by fission in the form of kinetic energy; the nuclei into which the atom splits (fission products) have very high speed; the fragments and particles are then slowed down by the particles collisions thus transforming their kinetic energy into thermal energy by friction. Figure 13.1 shows the nuclear fuel cycle.

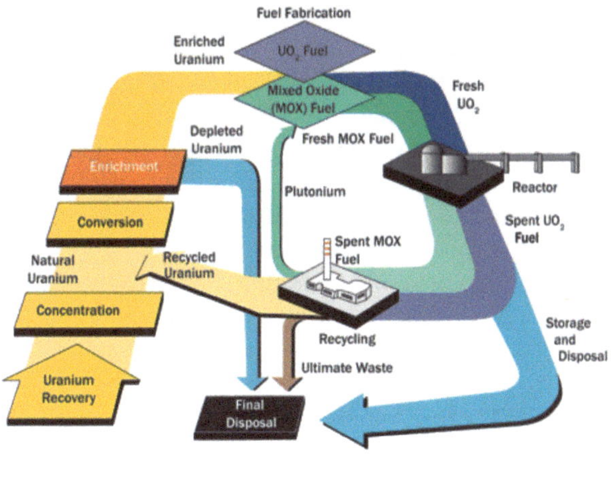

Fig. 13.1 Nuclear fuel cycle. *Source* infcis.iaea.org

In "thermal" reactors, neutrons are slowed down by a substance called a "moderator"; the "slow" neutrons have a greater probability of fissioning the nucleus of the fissile atoms, and the enrichment of Uranium can be low or even zero (natural Uranium). Moderators used in power reactors are graphite (natural Uranium can be used), heavy water (natural Uranium can be used), natural water (the Uranium must be enriched because H_2O captures too many neutrons). In "fast" reactors, neutrons, in the absence of a "moderator", have a much higher energy (speed); since the probability of fission is lower, higher enrichments in Uranium-235 are required (15–20%), but the probability of Uranium-238 fissioning becomes higher and there is also the capture of neutrons that produces fissile Plutonium: the produced fissile material is higher than what is consumed (hence the name self-fertilizers) (Fig. 13.1).

13.4 Nuclear Power Plants History

The first reactors were developed in the United States for the propulsion of submarines and aircraft carriers. In the United States in 1953 an experimental activity started on different types of nuclear reactors for generation of electricity, both pressurized (PWR) and boiling (BWR) water reactors, for which it was necessary to use Uranium enriched to at least to 3% in U-235, due to neutron capture by H_2O. The first reactors (prototypes) were built at the beginning of 1950s, while the first power plants, connected to the electricity grid, were built in the second half of the 1950s: in 1956 in Great Britain and in 1957 in the USA.

As regards France, in 1959 the first graphite gas reactor to produce electricity (50 MWe) came into operation. In the Soviet Union, in 1958, the first graphite moderated and boiling water-cooled reactor of 100 MWe came into operation, followed from 1959 to 1963 by another 5 units of the same power. Their name in Russian language is RBMK. The first commercial nuclear power plant in the USA came into operation in 1957, a PWR with a power of 50 MWe. In 1960 the first BWR came into operation in the USA, with a power of 200 MWe, followed from 1964 to 1971 by 8 BWRs with power from 23 to 660 MWe. In 1961 the second PWR came into operation, with a capacity of 175 MWe, followed from 1963 to 1968 by 3 other PWRs with power from 257 to 582 Mwe. The BWRs were developed by General Electric, while the PWRs were developed by Westinghouse and Babcock & Wilcox Companies. In 1964, the first liquid metal cooled fast neutron reactor, the 20 MWe Liquid Metal Fast Breeder Reactor (LMFBR), EBR-II, came into operation, and in 1966 the 66 MWe LMFBR Fermi-1 came into operation. In UK from 1960 to 1970, 16 gas-graphite power plants came into operation, with power from 50 to 490 MWe.

In the Soviet Union in 1964 the first Russian 265 MWe PWR came into operation, followed by a 366 MWe second unit in 1970. Their Russian name is VVER. As regard Italy, in 1964 the first (and last) graphite gas reactor (153 MWe) and the first 150 MWe BWR came into operation. In 1965 the first (and last) 260 MWe PWR came into operation. In the mid-1960s, Italy was the third country in the world in terms of installed nuclear power. In Japan the first (and last) Japanese 159 MWe graphite gas reactor came into operation in 1966. In Germany in 1966 the first German 70 MWe PWR came into operation. From 1967 to 1969, two 237 and 256 MWe BWR and a 340 MWe PWR came into operation. In France from 1964 to 1971 other 5 gas-graphite reactors came into operation and in 1967 the first 340 MWe PWR reactor came into operation; since then, France has built more than 50 PWRs. Concerning Canada, in 1968 the first 216 MWe heavy water reactor came into operation; it is the first of a large series of heavy water reactors, called CANDU (Canadian Deuterium Uranium). In Spain in 1969 the first 142 MWe BWR came into operation. In Holland in 1969 the first (and last) 55 MWe BWR came into operation.

Three main periods can be identified in the history of nuclear power plants:

(1) In the first period, from 1945 to 1954, prototypes for both marine engines and thermoelectric power plants were designed and built. This period was dominated by the policy of secrecy and the Anglo-American monopoly of Uranium.
(2) The second period, from 1954 to 1964, is characterized by the euphoria caused.by the suppression of secrecy, the entry into operation of the first power plants and the international control.
(3) The third period from 1964 to 1974 is that of industrial development, large electricity production programs and international competition.

After the third period there is a long period until today in which nuclear energy to produce electricity has reached industrial maturity. As operational experience increases (currently around 18 500 reactor years) nuclear power plants have become increasingly reliable, with high utilization factors (in the United States above 90%). After the first-generation nuclear power plants (power of 100–300 MWe), the second-generation power plants were developed between the 1960s and 1970s, with a power of 600–1000 MWe.

The second-generation nuclear power plants were built in the second half of the 1970s. For instance, in Italy the Caorso power plant (840 MWe BWR) came into operation in 1979. The construction of the Montalto di Castro power plant (2100 MWe BWRs) began in 1982, and the project of a 1000 MWe PWR was not accomplished, following the two Referendums of 1987 and 2011, after Chernobyl and Fukushima accidents, in which Italian people spoke out against the nuclear energy.

The 70s were the years of the greatest development in the world of nuclear power plants. In the 80s, however, a significant decrease in the construction of new power plants began: too optimistic forecasts were done, and the power plants are more complex plants than initially assumed. Among the reasons for the decline in the construction of new power plants are the three accidents at Three-Miles Island in the USA in 1979, at Chernobyl in 1986 in Ukraine, at Fukushima in 2011 in Japan, with the following consequences: less acceptability of populations and increasing costs to make the power plants safer. In 2022 there were around 440 nuclear power plants in the world (for a total electrical power of 396,000 MWe) in just thirty Countries. 53 Countries have a total of 220 research reactors, including medical-health ones (for example for the production of radiopharmaceuticals used in tumor therapies or tracer elements, such as Technetium, used in diagnostic imaging); 171 of these reactors were built with highly enriched Uranium fuel (HEU with more than 20%

Uranium-235), converted to reactors with lower enriched fuel (LEU) since 1978 (to reduce the risks of nuclear proliferation). Furthermore, there are around 350 nuclear submarines and dozens of American nuclear aircraft carriers (and some Russian icebreakers) in which nuclear energy is used for the naval propulsion.

13.5 Nuclear Power Plants Descriptions

Simplifying as much as possible, in a nuclear power plant there are: the so-called nuclear island, which includes the nuclear reactor and the primary coolant loop, the engine room, with the turbines and the electric generator (the so-called BOP: Balance Of Plant). In the PWRs the reactor includes a pressure vessel (up to around 160 bar), in which there is the reactor core, with the fuel rods and the control rods. The fuel rods are metal tubes, about 1 cm in diameter and about 4 m long, usually arranged vertically, containing Uranium oxide. The coolant flows between the rods and removes the thermal power; the primary coolant transfers heat to the secondary fluid in a heat exchanger (steam generator) and then the produced steam goes to the turbines (Rankine cycle). In the BWRs the steam is produced within the reactor core and flows to the engine room towards turbines. CANDU reactors do not have a pressure vessel, and the fuel bundles are inserted into pressurized horizontal pipes that are within the so called calandria, which is not pressurized.

Concerning the high temperature pebble bed gas cooled reactors (HTGR) at present there are reactors in China: the construction of a HTR-10 (a 10 MWt prototype pebble bed reactor) began in 1995 and it has been operating at full capacity since January 2003. Concerning the fast reactors, the most important fast breeder reactor, cooled with liquid sodium, was the 1200 Mwe Super Phénix in France (no longer in operation). Other Liquid Metal Fast Breeder reactors were built in various countries but have been shut down, except in Russia (BREST); BREST with an output power of 300 MWe, cooled with liquid lead, it is expected to start the operation in mid-2026. The NewCleo Company designed a reactor «small», cooled by liquid lead, self-fertilizer; the prototype is planned for 2030 and the first commercial reactor for 2032.

Once used, nuclear fuel (which constitutes a highly radioactive waste) is placed in a water pool adjacent to the reactor building that shields the radiation. After about two years the fuel is either placed in watertight metal and concrete containers, or chemically treated in such a way that only the vitrified residue needs to be stored in apposite casks. In both cases, the casks containing the waste must be placed in geological deposits at great depth (400–600 m), which must have suitable characteristics to safely preserve the radioactive wastes for thousands of years. The Onkalo geological repository (about 500 m deep) is a geological repository intended to permanently store Finnish radioactive wastes: it is the first repository in the world with these characteristics and it is under construction near the Olkiluoto nuclear power plant. In Sweden the chosen location was under Östhammar, 100 km north of Stockholm. The possibility of transmuting the radioactive waste to eliminate the longer-lived ones has been under study for some time: the radioactivity of most wastes lasts between 200 and 300 years; only small quantities, the so-called Transuranics, have a very long life. Plutonium, for example, has a half-life of approximately 24,000 years, it can be chemically extracted from the wastes (since it is fissile) and becomes nuclear fuel again, while other transuranic nuclei can be transmuted into short-lived nuclei.

13.6 Nuclear Accidents

The accident at Three-Miles Island-2 (Pennsylvania) in 1979 (Level 5 on the INES scale), due mainly to operator errors, caused the partial meltdown of the PWR core and the definitive shutdown of the reactor, with significant economic damage and the stop of the nuclear plants construction, even though there was no radioactive contamination in the surrounding area and no deaths. The Chernobyl accident (in Ukraine) in 1986 (Level 7 on the INES scale), caused by a human error in a plant with significant weak points, caused the destruction of the RBMK reactor, 65 immediate deaths, with forecasts of several thousand deaths in subsequent years, and serious consequences on the territory: all residents within a 30 km radius of the plant, approximately 116,000 people, have been relocated and the region remains uninhabitable. The Fukushima accident in 2011 in Japan (Level 7 on the INES scale), after the 9.0 magnitude earthquake followed by a tsunami, caused the destruction of 4 BWRs, the contamination of a vast territory and 110,000 people evacuated, the definitive closure of six reactors, enormous economic costs (even if there were no deaths during the accident).

13.7 New Lines of Development

In the last decade, some so-called third generation reactor plants have been built (such as American Westinghouse AP1000, French European Pressurized Reactor EPR, Russian VVER1200, Chinese Hualong 1, Korean APR1400), with enhanced safety systems based on passive systems, which do not require energy (neither internal nor external) to start and be effective.

Concerning Generation IV, nuclear reactors are being developed through the international cooperation of 14 countries, with construction planned after 2030. The Generation IV International Forum (GIF) was promoted by the U.S. Department of Energy in 2000 and formally established in mid-2001. It is an international group including the governments of 13 countries where nuclear energy is significant now and considered vital for the future. Most are committed to jointly developing the next generation of nuclear technology.

The original founding members of GIF are Argentina, Brazil, Canada, South Korea, France, Japan, South Africa, the United Kingdom and the United States. They were joined by Switzerland, China, Russia, Australia and the European Union.

The aim is that these innovative systems must be cleaner, safer and more efficient than previous generations, with minimization and non-proliferation of radioactive wastes (e.g. the use of Uranium–Plutonium Mixed Oxide fuel containing minor Actinides in the long term makes it completely ineffective to produce Plutonium for military uses). In fast breeder reactors, high-activity, long-lived wastes are burned (transmuted) much better than in thermal reactors, with reduction of the source of potential radiotoxicity in a geological repository and increase in the capacity of the geological repository. Six Generation IV nuclear systems were selected and studied: Sodium Fast Reactor, Lead Fast Reactor, Gas (Helium) Fast Reactor, Molten Salt Reactor (all with closed fuel cycle and breeders), Very High Temperature Reactor (Helium-cooled), Super critical Water Reactor. The Global Nuclear Energy Partnership (GNEP) was signed in 2007 by 20 countries (USA, China, France, Japan, Russia, Australia, Bulgaria, Ghana, Hungary, Jordan, Kazakhstan, Lithuania, Poland, Romania, Slovenia, Ukraine, Italy, Canada, South Korea, Senegal). It intends to develop consensus throughout the world on the use of nuclear fuel that allows all countries to obtain it, avoids non-proliferation and distorted use, and minimizes wastes.

Innovative fuel reprocessing processes are being studied: if the fuel is reprocessed after the use (closed cycle), the radioactive waste is « vitrified». This operation significantly reduces storage space to approximately 1/6 compared to direct storage. More recently, and to address the problem of high capital costs, innovative small modular reactors (SMRs) have been studied, starting in the United States, and could be built within the next decade. These innovative reactors, ranging in size from a few MWe to a few hundred MWe, can be used for power generation, processing heat, desalination or other industrial uses. SMRs can use light water as a coolant or other coolants such as gas, liquid metal or molten salt. Advanced SMRs are characterized by relatively small size and low capital investment. They can be built in locations that are not possible for large nuclear power plants and are suitable for incremental power additions (hence Modular). SMRs also offer significant benefits in terms of safety, security and non-proliferation issues. The American NuScale reactor (77 MWe, according to the last design) is an example of an SMR: if necessary, the NuScale SMR shuts down and cools itself for an indefinite period, without requiring any action from the operator, without water additional and without AC or DC power. The NuScale power module's expected cost per KWh is competitive with other baseload electricity generation sources. Figure 13.2 presents the NuScale reactor, showing the passive decay heat removal system and the passive containment heat removal system. In Fig. 13.3 the reactor building, with several modules, is shown (Figs. 13.2 and 13.3).

The Traveling Wave Reactor (TWR) is a visionary projects by TerraPower (founded by Bill Gates), which uses a fuel cycle without recycling: TWR starts the reaction with enriched Uranium, then switches to depleted Uranium coming from enrichment plants. A TWR reactor could run on depleted Uranium for about 100 years. It does not require fuel reprocessing, and has many attractive features (passive safety, waste reduction, reduction and eventual elimination of enrichment, and high thermal efficiency). The Canadian Nuclear Safety Commission (CNSC)

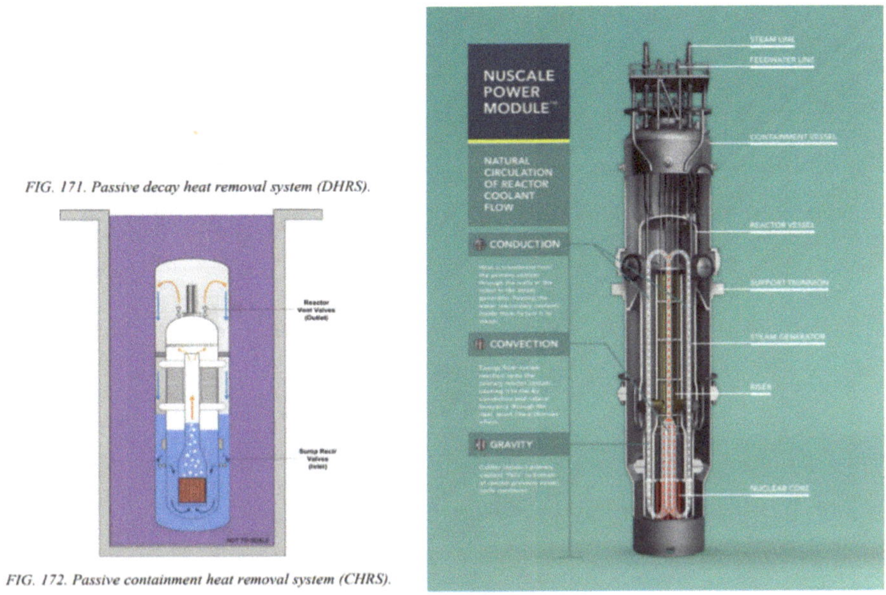

Fig. 13.2 Nuscale reactor. *Source* https://www.nuscalepower.com

Fig. 13.3 Nuscale reactor building with several modules. *Source* https://www.nuscalepower.com

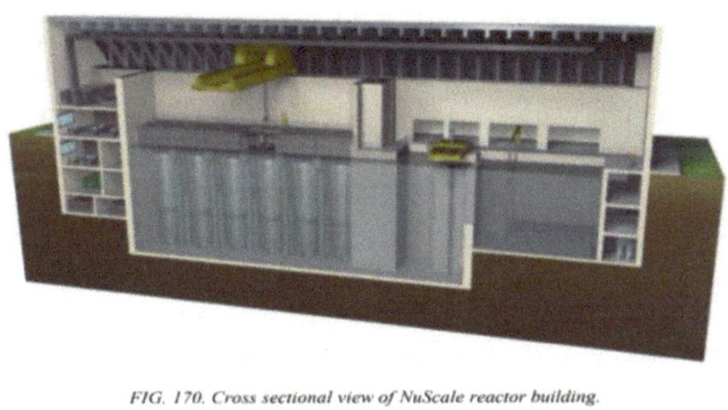

and the United States Nuclear Regulatory Commission (NRC) have selected Terrestrial Energy's Integral Molten Salt Reactor (IMSR), shown in Fig. 13.4, for the first joint technical review of advanced nuclear reactor technology, cooled with molten salts. The IMSR integrates the primary reactor components, including the primary heat exchangers with the secondary non-radioactive molten salt loop, into a sealed replaceable central vessel that has an expected life of seven years. It operates at 600–700 °C and can support many industrial process heating applications. The reactor uses a graphite moderator; the molten salt fuel is an eutectic fluoride with low Uranium fuel at atmospheric pressure. Emergency cooling and heat removal are based on passive systems (Fig. 13.4).

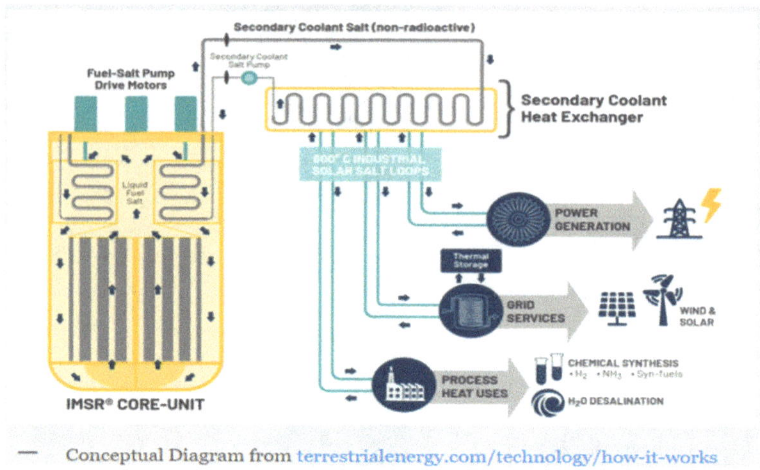

Fig. 13.4 IMSR. *Source* https://www.terrestrialenergy.com

The American company Oklo announced in December 2019 an innovative nuclear reactor that operates on a single fuel load for decades: Aurora Advanced Fission. Calling it a "clean energy facility," the company says the power plant would be integrated with solar panels to provide communities with reliable electricity 24 × 7, 365 days/year. The reactor is designed to produce approximately 1.5 MWe, having the capacity to produce process heat for residential or commercial applications.

The Innovative Heavy Water Reactor (AHWR) is a 300 MWe boiling light water cooled, heavy water moderated, vertical pressure tube reactor, designed to produce most of its power with Thorium, having a desalination capacity of 500 m^3/day. Nuclear fuel consists of Mixed Oxides of Th-U233 and Th-Pu. The plant flow sheet is shown in Fig. 13.5.

SMART (System Integrated Modular Advanced Reactor) is a 330 MW PWR being studied at KAERI (Korea Atomic Energy Research Institute). The reactor will be used as an energy source for seawater desalination and small-scale power generation. It is provided with intrinsic and passive safety features. SMART (Fig. 13.6) is designed to provide 40,000 tons of fresh water per day and 90 MWe in an area with a population of 100,000 inhabitants (Fig. 13.6).

ACP100 was developed by China National Nuclear Corporation (CNNC): it is an integral PWR with a power of 100 MWe. The reactor can be used for electricity generation, process heat and seawater desalination.

From an idea by Carlo Rubbia (Nobel Prize winner in 1984), the ADS reactor is an energy-amplifying nuclear reactor supported by an external source of protons (Accelerator-Driven System or ADS, shown in Fig. 13.7), necessary to power the nuclear reaction in the core of the reactor. This is a subcritical reactor, not able to independently carry out a chain reaction. To trigger the chain reaction, it needs to be irradiated by a powerful beam of particles (protons) coming from an accelerator. The reactor releases sufficient thermal energy to be transformed into electrical energy

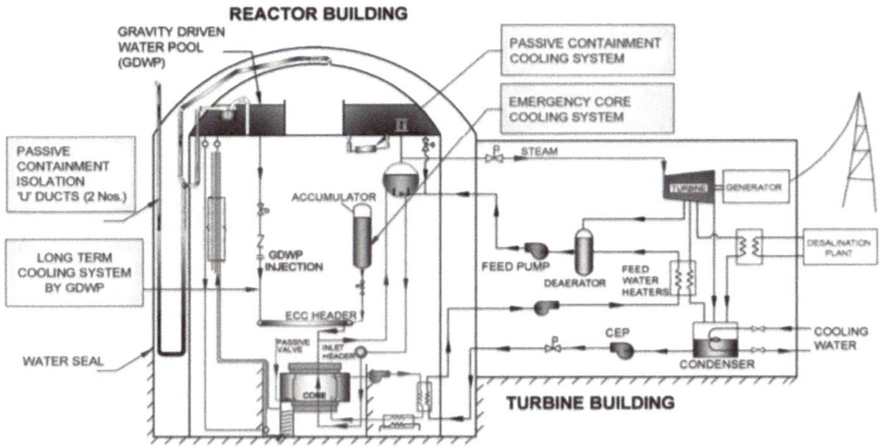

FIG. 165. Simplified flow sheet of AHWR.

Fig. 13.5 AHWR simplified flow sheet. *Source* https://www.barc.gov.in/randd/ahwr.html

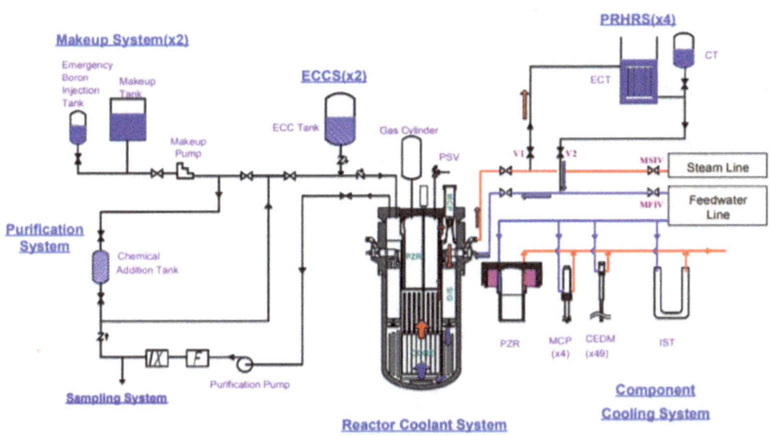

FIG. 169. Schematic diagram of the SMART NSSS.

Fig. 13.6 Smart scheme. *Source* https://www.kepco-enc.com

which will be used to power the accelerator and provide a surplus of electrical energy for the electricity grid (Fig. 13.7).

MYRRHA is a multipurpose hybrid research reactor for high-tech applications: it is a "first of its kind" design of a nuclear reactor coupled to a proton accelerator (an accelerator-driven system—ADS). MYRRHA is a lead–bismuth-cooled fast reactor with two possible configurations: subcritical or critical. The reactor is built at the SCK • CEN, the Belgian Center for Nuclear Research, and has already been

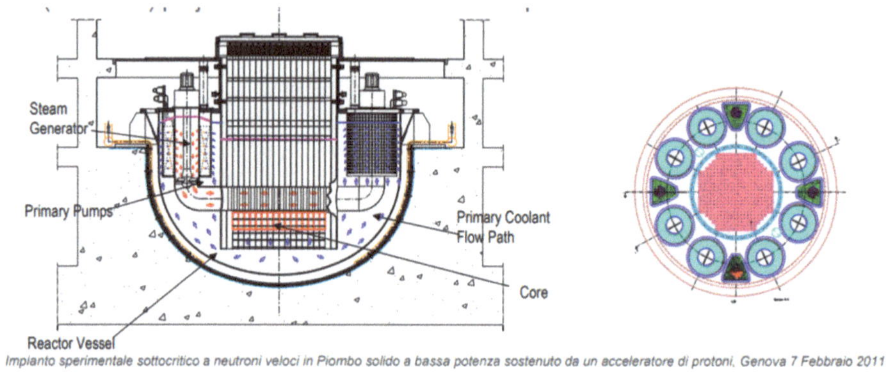

Fig. 13.7 ADS. *Source* https://world-nuclear.org

funded by the Belgian Government. MYRRHA (shown in Fig. 13.8) is expected to be fully operational in 2036, with a first phase (100 MeV accelerator) ready in 2026 (Fig. 13.8).

A Report by IAEA Advanced Reactors Information System (ARIS), "Advances in Small Modular Reactor Technology Developments: A Supplement to 2020 Edition", illustrates the characteristics and level of development of 72 modular reactors of up to 300 MWe power. The USA Energy Secretary recently announced Small Modular Reactor funding, six contracts for high-assay low -enriched Uranium (HALEU) deconversion and four contracts for HALEU enrichment services. Under these contracts, "selected companies will bid on future work to produce and store HALEU

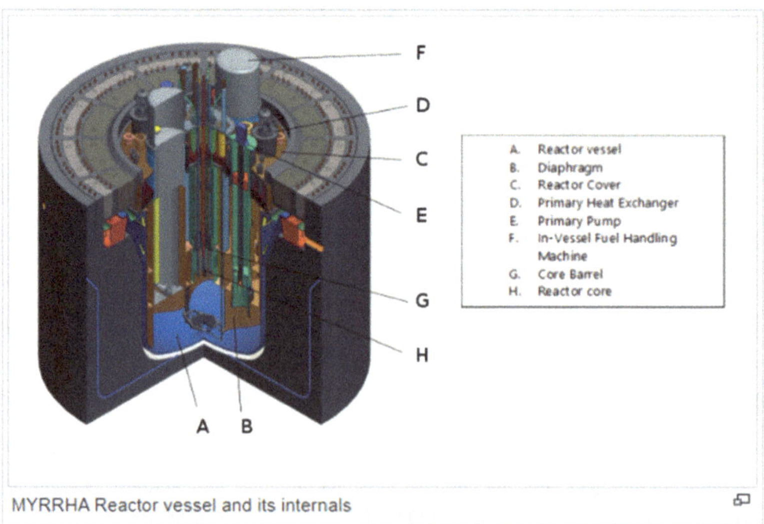

Fig. 13.8 MYRRHA nuclear plant. *Source* https://www.sckcen.be

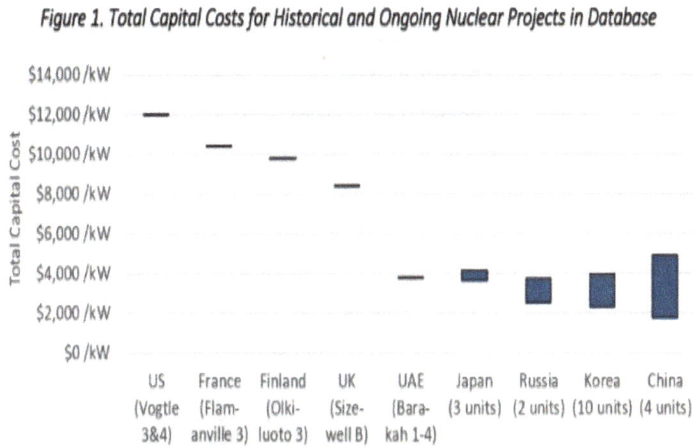

Fig. 13.9 Capital costs in several countries. *Source* https://world-nuclear.org/information-library/economic-aspects/economics-of-nuclear-power

in the form of Uranium hexafluoride gas to eventually be made into fuel for advanced reactors".

Concerning the nuclear energy cost a significant problem in OECD countries is the high capital costs (particularly after the Fukushima accident). There is also a big difference in capital costs between the US, France, Finland and the UK on one side and Russia, Korea and China on the other side (for instance $12,000/kW in the USA versus $5000/kW in China), as it is shown in Fig. 13.9.

The contributions to the cost of nuclear energy are shown in Fig. 13.10.

The nuclear energy costs depend also on the construction duration. According to data collected by the IAEA on the 61 new nuclear reactors connected to the grid over the past decade, units in the Far East were built almost twice as fast as those in Europe (66 months versus 110 months).

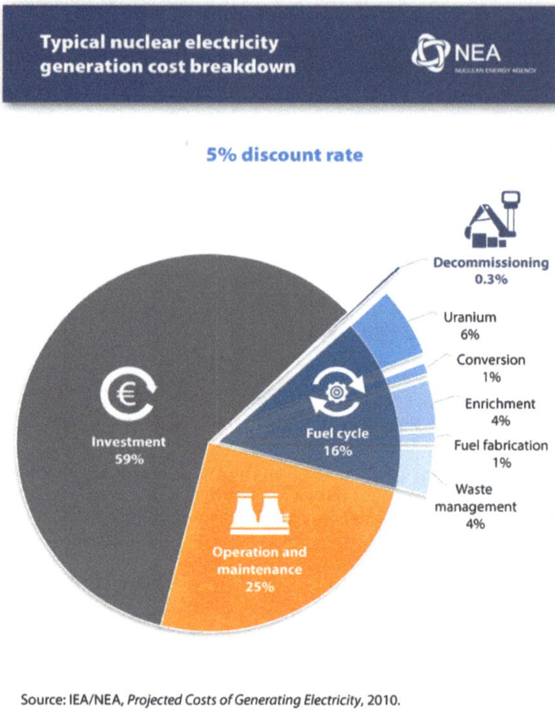

Fig. 13.10 Nuclear KWh cost contribution

13.8 Conclusion

There is an increasing interest in nuclear energy worldwide, particularly for the Small Modular Reactors (for instance those designed by Nuscale and NewCleo), but the prototypes have to be still tested. In the medium term (2040–2050) a strong increase of nuclear power plants energy in France, China, Russia and India (with the export of nuclear reactors in various Countries) is very likely. In Italy, despite the referendum against the nuclear energy, there is again the hope to build some small modular reactors; there are also some industries and research centers, like Ansaldo nucleare in Genova and ENEA, as well some nuclear engineering departments in different Universities, from which several hundred graduates emerge.

Chapter 14
The Non-proliferation Treaty at Its Midlife Crisis

Alessandro Pascolini

Abstract Half a century ago, the Treaty on the Proliferation of Nuclear Weapons (NPT) came into force, one of the fundamental international instruments for global order and security. It established the rules for the safe development of peaceful nuclear technologies and became the bedrock for the intricate structures of nuclear arms control that made possible the sharp reduction of the American and Russian nuclear arsenals during the 1990s. Over the years the NPT has scored significant successes, but also experienced delicate vicissitudes, which are growing in the last few years; the dramatic present global situation and recent developments show signals threatening its very continuity. Because the treaty is part of a larger context of different disarmament and non-proliferation initiatives, its success or failure is closely intertwined to the developments of all these initiatives, and the global security of the whole world. The Non-Proliferation Treaty is going through its 54th year, a delicate age for human beings, exposed to the risk of a midlife crisis, a period of transition in life where someone struggles with his identity and self-confidence. At this critical point, it is worth remind the original spirit and aims of the treaty.

14.1 Towards the Treaty

At the beginning the tenet was: non-proliferation is a good in itself. From the end of the 1950s several resolutions of the United Nations General Assembly (UNGA) expressed need for a nuclear non-proliferation regime to avert the risks to peace and security from the increase in the number of states possessing atomic weapons, also making the process of reducing existing weapons more difficult. This is the spirit of Resolution 1665, presented in 1961 by Ireland and unanimously approved.

A marked change in the terms of the proliferation debate occurred in 1965 with the intervention of not aligned countries: non-proliferation is a potential obstacle to the national security of smaller countries [1], each state had the right to its own

A. Pascolini (✉)
University of Padova and SICA—International Security and Arms Control Committee, Rome, Italy
e-mail: alessandro.pascolini@pd.infn.it

P. Cotta-Ramusino et al. (eds.), *Nuclear Weapon Risks in a Problematic Time*, Springer Proceedings in Physics 358, https://doi.org/10.1007/978-3-032-29873-7_14

nuclear armament, which could only be renounced in return for compensation from the nuclear powers.

This led to the UNGA Resolution 2028 (November 1965), which indicated to the 18-Nation Committee for Disarmament (ENDC) as a principle for a future treaty: to generate a satisfactory balance of responsibilities and reciprocal obligations between nuclear-weapon states and non-nuclear-weapon states; to be an instrument towards the achievement of general and complete disarmament and more specifically nuclear disarmament.

In the following years both the US and the Soviet Union presented the UNGA and the ENDC with various draft treaties to limit the spread of nuclear weapons. The point of greatest divergence between the two states concerned the admissibility of joint management of the nuclear arsenal within a regional organization, a position supported by the USA, which wanted to create a multilateral nuclear force in NATO, and strongly opposed by the USSR, which could not accept a nuclear-armed Germany.

A further condition emerged in the negotiations, concerning access to civilian applications by non-nuclear countries, which wanted to avoid remaining in a position of perpetual inferiority in critical nuclear technologies.

The impasse in the negotiations between the USA and the USSR was overcome by the American renouncement of the creation of a multilateral nuclear force in NATO and the Soviet acceptance of American bases in Europe: as a matter of fact the treaty does not explicitly prohibit nuclear powers from stationing nuclear weapons in non-nuclear countries.

On 18 January 1968, the USA and the Soviet Union presented a joint draft treaty to the ENDC; after further discussion and revisions [13], UNGA Resolution 2373, with the text of the treaty as an annex, passed on 12 June with 95 votes in favor, 4 against and 21 abstentions. On 19 June 1968, the Security Council (UNSC), in order to meet the concerns expressed by the non-aligned countries, passed Resolution 255 on Positive Security Assurances (PSA) for the non-nuclear NPT countries: in the event that they were the subject of an attack or threat of a nuclear attack, the permanent members of the UNSC would have to take the necessary measures to counter it.

The treaty opened for signature on 1 July 1968 and came into force on 5 March 1970 with the ratifications of the depository countries (UK, USSR and USA) and of 40 other countries [3].

14.2 Treaty's Objectives and Dictates

The treaty sets the terms of the triple 'contract' between non-proliferation, promotion of peaceful nuclear applications and nuclear disarmament [16, 17].The preamble highlights the crucial political terms and objectives of the treaty:

- make every effort to prevent the danger of nuclear war,

- prevent the proliferation of nuclear weapons, which would increase the danger of nuclear war,
- ensure the right of every country to the benefits of peaceful applications of nuclear technology, and to the fullest exchange of scientific information for the development of peaceful purposes,
- strengthen the crucial role of the safeguards on civilian activities by the International Atomic Energy Agency (IAEA),
- commit to negotiate the prohibition of all nuclear explosive tests,
- achieve a halt to the nuclear arms race as soon as possible and take effective steps for nuclear disarmament, up to the elimination of all nuclear weapons and their means of delivery following a comprehensive disarmament treaty,
- refrain states from the threat or use of force against the territorial integrity or political independence of any state,
- promote the establishment and maintenance of international peace and security.

The NPT is *unique* among international legal instruments in that it admits an inequality of rights and obligations between the parties: on the one hand the states that detonated an explosive device before 1 January 1967 (*Nuclear-weapon states*—NWS, i.e. China, France, UK, USSR and USA), and on the other hand the remaining countries (*Non-nuclear-weapon states* –NNWS). In principle, this discrimination is only temporary, i.e. until the nuclear disarmament is achieved, but in fact it is being perpetuated over time.

The NNWS accepted this asymmetry by realistically recognising that the NWS were not prepared to give up their weapons any time soon and that in any case the overall compromise of the treaty was beneficial because of the NWS's commitment to minimize the negative political economic and security consequences of NNWS' renounce. Additionally, the NPT improves national security of individual NNWS as it ensures that neighboring nations cannot develop nuclear weapons and use them for coercive actions.

The first two articles concretise non-proliferation: the NWS undertake not to transfer nuclear weapons to others (Article I), while the NNWS renounce all military nuclear developments (Article II).

Articles III, IV and V deal with the peaceful development of nuclear energy and its control, introducing IAEA's safeguards on NNWS' facilities and fissile materials to prevent diversion from peaceful uses; all transfers of nuclear materials and technology to NNWS are subjected to IAEA control. The controls must be compatible with the "the inalienable right of all Parties to develop research, production and use of nuclear energy for peaceful purposes without discrimination" (Article IV).

The lapidary (and largely ambiguous) Article VI expresses the prospects for disarmament:

> Each of the Parties to the Treaty undertakes to conduct negotiations as soon as possible in good faith on effective measures relating to the cessation of the nuclear arms race and nuclear disarmament, and on a comprehensive and complete disarmament treaty under strict and effective international control.

Article VII recognises the right of any group of states to establish nuclear-free zones (NWFZs). The last four articles contain procedural clauses. Article X states that each party shall have the right to withdraw from the treaty in case of extraordinary events endangering the supreme interests of the country; for the treaty itself, a conference is convened 25 years after entry into force to decide, by majority vote, on its extension indefinitely or for defined periods of time.

It should be noted that the NPT is not self-fulfilling and 'external' interventions are required for the implementation of each point, the IAEA for Art III, inter-state agreements for Art IV, specific treaties for the NWFZs, the Conference on Disarmament or UNGA for Art VI.

The NPT has contributed significantly to the creation and sustainment of the global non-proliferation regime: adherence to the treaty has led to the final closure of the military nuclear option in the adhering countries, facilitated the creation of nuclear weapon-free zones, which now span the entire southern hemisphere [16, 17], slowed down the nuclear projects of non-adhering countries with its restrictions, and created the conditions for the unilateral or agreed downsizing of the military arsenals of the NWS.

Over the years, countries became convinced of the treaty's importance and the numerous accessions in the five-year period 1990–1995 made the NPT nearly universal [18], India, Israel, Pakistan and South Sudan never joined the treaty, and North Korea (DPRK), which acceded in 1985, withdrew from it in 2003. Taiwan signed the NPT in 1968 as the 'Republic of China' and remained a party to it until its expulsion from the UN in 1971; it has since continued to abide by its terms under a trilateral agreement with the IAEA and the US.

14.3 Safeguards and Breaches

To meet the conditions set by the NPT, the IAEA updated its control system in operation since 1961 in the Comprehensive Safeguards Agreement (CSA), which commits the state to accept safeguards in all peaceful nuclear activities and on all fissile materials. The IAEA monitors only declared nuclear facilities, examines the information provided by states, and conducts scheduled inspection visits, with the objective of detecting the diversion of a quantity of fissile material capable of producing an explosion (IAEA 1998).

Due to the inherent difficulty of separating civilian and military nuclear activities, pre-CSA control measures proved insufficient to prevent violations of the treaty and the establishment of a true nuclear black market by Pakistani scientist Abdul Qadeer Khan [7]. The main breaches regard Iraq, Libya, South Korea, Iran [11] and the DPRK, both when part of the treaty and on her withdraw. The NWS's resistance to substantive compliance with Article VI is largely considered to be a serious permanent violation of the treaty.

The need for a stronger system of safeguards was reinforced by the revelation of Iraq's and South Africa's nuclear programmes, which had escaped controls, and by the difficulties encountered in applying the CSA to North Korea. In 1997 the IAEA Board of Governors approved the *Additional Protocol*, which strengthens the control and safeguards measures [5].

The enhanced safeguards of the Additional Protocol aim at a holistic approach to a country's nuclear programme, with more powerful means of investigation and greater scope for action, including the use of advanced analysis technologies. In particular, it allows country-specific safeguard programmes to be defined. Today, Additional Protocols are in force with 151 States and Euratom. Another 13 States have signed the Additional Protocol but have yet to bring it into force.

14.4 NPT Review Conferences

Given the NPT's crucial role for global nuclear security, a conference is scheduled every five years to "review the operation of the Treaty to ascertain whether the purposes of its preamble and its provisions are being fulfilled" and to make suggestions for strengthening the control of military and civilian nuclear energy.

The conferences measure the life-evolution of the treaty and reflect the global international climate. Most of them have been characterized by the clash between NWS and NNWS, the former insisting on stricter conditions and greater guarantees of non-proliferation, the latter recalling NWS' obligations regarding nuclear disarmament and the commitment to non-nuclear countries' security guarantees, both positive (PSA) and negative (NSA); another frequent NNWS complaint is the discriminatory limits on exports of civilian nuclear materials and technologies imposed outside the NPT by the *Zangger Committee* and the *Nuclear Suppliers Group* of producer states.

States participate in the conferences individually and as members of groups with a single spokesperson, based on continental groupings, or formed on the basis of specific disarmament initiatives. Recently, accredited non-governmental organisations also take part in conferences, where they can present their own papers.

Since in these conferences every decision, including the agenda of the proceedings, requires unanimity of consensus, the adoption of a final document is not always possible [2], especially on the occasion of serious political tensions,moreover, as is often the case in large international assemblies, when a general consensus has to be achieved, some of the more incisive proposals are scaled down and hot topics are avoided. So far, only in 1975, 1985, 1995, 2000 and 2010 has unanimity been reached on a final document, albeit often with difficulty and setting aside key issues.

Of critical importance was the 1995 Review and Extension Conference (Art. X(2)), the first in which all five NWS participated. Some states proposed extending the treaty for defined periods subject to review, fearing that an NPT without time limits would 'mummify' the asymmetry between NWS and NNWS. The vast majority was for the indefinite extension, and so it was decided, once certain conditions set by NNWS and in particular the Arab countries were accepted, including the establishment in the Middle East of a zone "free of nuclear weapons and all other weapons of mass destruction" (MEWMDFZ).

The 2000 Review Conference set out specific disarmament commitments in 13 "steps", including the ratification of the Comprehensive Nuclear-Test-Ban Treaty (CTBT) and a moratorium on nuclear tests, the conclusion of a fissile material cut-off *treaty (FMCT)* and regular reports on Article VI compliance measures [15]. None of the 13 points were adequately realised in the following five years, and the negative developments in international politics during this period turned the 2005 conference into a continuous confrontation: two weeks were needed just to set the agenda for the work, which then did not lead to any significant results [9].

Barak Obama's new international policy recreated trust for the non-proliferation regime [14] and the eighth conference concluded in May 2010 with a final document, unanimously approved. On nuclear disarmament, a plan of 22 operational actions was identified, with effective initiatives for nuclear security and counter-terrorism, and 42 actions (not particularly strong) were approved to strengthen non-proliferation.[1]

A crucial topic of the conference was the Middle-east Nuclear free zone: a condition for the adoption of the final document was the commitment to hold a conference on the topic in 2012 with the participation of all countries in the area. The appointed facilitator was unable to overcome the divergent positions of Israel and the Arab countries, and the failure to convene the conference[2] was the formal cause for the failure of the 2015 review conference, which witnessed multiple irreconcilable disagreements and irreducible positions on the majority of the points under consideration [10]. The NWS hardened their positions and argued that nuclear disarmament must be 'based on the principle of global strategic security and stability' and downplayed the disarmament obligations expressed in 2010 as mere 'recommendations' rather than binding commitments. Even on the non-proliferation side, there were no points of contact and thus no strengthening of the regime.

A further radical confrontation between states that consider nuclear weapons as legitimate security instruments and those who want their abolition arose with the Humanitarian Impact Initiative [8], elaborated and consolidated in the international conferences of Oslo (2013), Nayarit and Vienna (214), which then evolved into a real negotiation outside the NPT, leading to the *Treaty on the Prohibition of Nuclear Weapons* (TPNW) approved by 122 countries on 7 July 2017.

The 10th review conference, originally scheduled for 2020 and repeatedly postponed due to the COVID-19 pandemic, convened on 2022 at an extraordinarily difficult international environment. Russia's invasion of Ukraine and the safety of the Zaporizhzhia nuclear power plant exacerbated concerns about the risk of a nuclear accident. The war was of direct relevance to treaty implementation and the review conference deliberations, from the violation of security assurances provided to

---

[1] 2010 Review Conference of the Parties to the Treaty on the Non-Proliferation of Nuclear Weapons. Review of the operation of the Treaty, Conclusions and recommendations for follow-on actions, Final Document Volume I, Part I, NPT/CONF.2010/50 (Vol. I). New York: United Nations.

[2] Russia UK and US (2015) Middle East Nuclear and Weapons of Mass Destruction Free Zone: progress towards the convening of a Conference attended by all States of the Middle East, NPT/CONF.2015/WP.48, 1 May 2015. New York: United Nations.

Ukraine when it acceded to the NPT (*Budapest Memorandum*) to the safety and security of nuclear facilities and the ability of the International Atomic Energy Agency to continue implementing safeguards in occupied Ukrainian facilities.

Widely expected to be a disaster, the conference came surprisingly close to adopting a final document [12]. Although most delegations were disappointed with the draft outcome document, which was short on forward-looking disarmament steps, all the states-parties, except Russia, were prepared to join the consensus in an apparent effort to shore up the NPT regime, which has not had an agreed outcome in more than a decade. In the end, however, the NPT review conference failed over the impossibility of reconciling the positions on the war against Ukraine, and Russia opposed the adoption of the report.

14.5 Towards the 11th Review Conference Scheduled on 2026

Given the high political and diplomatic value of review conferences and the complexity of the work to be done, it was decided in 2000 to engage the international community in the three years preceding a RevCon in preparatory work with a three-session preparatory committee.

The first prepcom for the 11th Conference was held in August 2023 and, according to observers, resulted in two weeks of deplorable diplomacy, lack of leadership and the dominance of narrow national interests that paralyzed the work on all the issues being discussed. It ended with unprecedented procedural struggles reflecting very serious political disagreements.[3]

The second Prepcom took place from 22 July to 2 August this year. The committee found itself at a time of serious global tension, with significant developments in nuclear issues, including: the modernization of the nuclear arsenals of all NWS and the rise of Chinese weapons, the suspension of the bilateral *New Start* treaty, Russian nuclear threats, the placement of Russian nuclear weapons in Belarus, the Australian acquisition of nuclear-powered submarines (AUKUS agreement), North Korea's new tactical nuclear weapons, Iran's production of highly enriched uranium, Russia's military occupation of the Ukrainian Zaporizhzhia power plant and resulting security concerns, NATO's heavy focus on its nuclear capabilities, the resumption of nuclear test technology, and programs for intermediate-range missiles in the Indo-Pacific and Eurasia arenas.

New complexities and uncertainties were also considered in the rapidly evolving intersections of nuclear weapons, new technologies and new domains of conflict, outer space and cyber space, applications of artificial intelligence, missile defences and hypersonic launchers. From the doctrinal point of view, there have been lively

[3] Preparatory Committee for the 2026 Review Conference of the Parties to the Treaty on the Non-Proliferation of Nuclear Weapons. First session. (2023). Report of the Preparatory Committee on its first session. NPT/CONF.2026/PC.I/6. New York: United Nations

discussions about the principle of nuclear deterrence itself, its compatibility with disarmament, and the serious risks involved.

Discussing proliferation issues, the focus was on the forms of safeguards provided by IAEA and the desirability of universal adherence to comprehensive safeguards agreement.

In the discussion of peaceful applications of nuclear energy, a novelty was the presentation by several countries of the risks and dangers posed by electronuclear facilities facilities (including small modular reactors-SMRs and 'advanced' facilities) to people and the planet with the suggestion that they be phased out in favour of environmentally friendly sources. These perspectives were reinforced in the discussion of the threat of catastrophe at the Zaporizhzhya nuclear power plant and Japan's plan to discharge tritiated waste water from the Fukushima Daiichi nuclear power plant into the ocean.[4]

One session was devoted to strengthening the NPT and making the work of the RevComs more effective, but any operational proposals, particularly for transparency measures for NWS nuclear forces, were vetoed by Russia and China.

After discussion of the draft summary, the chairman noted that the document did not enjoy consensus.[5] Even on the hitherto uncontroversial "importance of promoting the participation of women in the process of nuclear non proliferation" (item 52) there were Iran's reservations. The outlook for the 11th Review Conference is far from rosy.

14.6 Looking at the Health Status of the NPT

According to Encyclopaedia Britannica [4], middle age is a period of adjustment between the potentialities of the past and the limitations of the future. Individuals accept the limits of their accomplishments and either take satisfaction in them or despair and become anxious over unachieved goals or unrealized aspirations.

We can then look at the main aims of the NPT as indicated in its preamble to check for the symptoms of a possible midlife syndrome.

About proliferation: five additional countries have produced nuclear weapons since 1970 and we have two threshold states, Iran and Japan, which can produce a nuclear armament in a period of months should the decision be taken. They both have:

- the necessary fissile material (low-enriched uranium or separated plutonium),

---

[4] Preparatory Committee for the 2026 Review Conference of the Parties to the Treaty on the Non-Proliferation of Nuclear Weapons. Second session. (2024). Revised Draft Factual Summary. NPT/CONF.2026/PC.II/CRP.2/Rev.1. New York: United Nations.

[5] Preparatory Committee for the 2026 Review Conference of the Parties to the Treaty on the Non-Proliferation of Nuclear Weapons. Second session. (2024). Towards 2026 Reflections of the Chair of the 2024 session of the Preparatory Committee. NPT/CONF.2026/PC.II/WP.43. New York: United Nations.

- the means and expertise to enhance it to explosive quality,
- all the precursors necessary (manufacturing, expertise) to fashion a nuclear device,
- delivery vehicles, as ballistic missiles of several ranges.
- On the political side, we witness grave signals:
- In January 2023, South Korean President YoonSuk-yeol floated the idea that Seoul might need its own nuclear force to counter North Korea's growing strategic arsenal.
- In September 2023, Saudi Crown Prince Mohammed bin Salman repeated a well-known proliferation threat, telling Fox News that if Iran acquired the bomb, Saudi Arabia would also "have to get one".
- German politicians and scholars increasingly speak of a crucial Zeitenwende (turning point) with respect to non-proliferation.
- In Japan itself important politicians are not excluding nuclear developments.

The crucial issue remains the disregard for progress in nuclear disarmament, which has seen only backward steps, with the NWS determined to support the current phase of substantial nuclear arms build-up (including new kinds of weapons and extensions to the outer space and cyber domains) as necessary "given the international security environment."

Recalling that Article VI also calls for a "comprehensive and complete disarmament treaty", it should be noted that the world is moving in the opposite direction, with. a global rearmament of conventional forces and record highs in military spending, creating strong incentives for further nuclear proliferation. The entire arms control architecture is under stress, including the biological and chemical weapons conventions, and there are no prospects for a resumption of negotiations on nuclear weapons. A novel feature is also making nuclear arms control more difficult. Historically, the area of inter-state relations pertaining to nuclear weapons remained outside the contemporary political unrest. Russia and the United States kept nuclear weapons '*apart*' from current policy, and their armed conflicts did not prevent arms control negotiations. Now the war in Ukraine has marked the end of this de-coupling, crucial for actual arms control.

And, finally, NPT is realizing that it can hardly compete with the younger, smarter and less demanding *Treaty on the Prohibition of Nuclear Weapons*, which is attracting a growing number of countries parts of NPT.

I leave the prognosis to the readers.

References

1. P.C. Bleek, Why do states proliferate?, in *Forecasting Proliferation in the 21st Century: the Role of Theory*. ed. by W.C. Potter, G. Mukhatzhanova (Stanford University Press, Palo Alto, 2010), pp. 159–192
2. J. Dhanapala, T. Rauf, *Reflections on the Treaty on the Non-Proliferation of Nuclear Weapons: Review Conferences and the Future of the NPT* (SIPRI, Stockholm, 2017)

3. J. Dhanapala, R. Rydell, Multilateral diplomacy and the NPT: an insider's account, in *UNIDIR/ 2005/3* (Geneva: United Nations, 2005)
4. Encyclopaedia Britannica, Middle age; psychology (2024). https://www.britannica.com/science/middle-age
5. International Atomic Energy Agency, *Model protocol additional to the agreement(s) between state(s) and the International Atomic Energy Agency for the application of safeguards, INFCIRC/540* (IAEA, Vienna, 1997)
6. International Atomic Energy Agency, The evolution of IAEA safeguards, in *International Nuclear Verification Series No.2* (IAEA, Vienna, 1998)
7. S.N. Kile, Nuclear arms control and non-proliferation, in *SIPRI Yearbook 2005* (Oxford University Press, Oxford, 2005), pp. 551–577
8. A. Kmentt, The development of the international initiative on the humanitarian impact of nuclear weapons and its effect on the nuclear weapons debate. Int. Rev. Red Cross **97**(899), 681–709 (2015)
9. R. Johnson, Politics and protection: why the 2005 NPT review conference failed, in *Disarmament Diplomacy 80 (Autumn)* (2005), pp. 3–32
10. O. Meier, *The 2015 NPT Review Conference Failure—Implications For the Nuclear Order* (Institute for International and Security Affairs German, Berlin, 2015)
11. G. Mukhatzhanova, Pride and prejudice: understanding Iran's nuclear program, in *Forecasting Proliferation in the 21st Century: a Comparative Perspective.* ed. by W.C. Potter, G. Mukhatzhanova (Stanford University Press, Palo Alto, 2010), pp. 42–75
12. G. Mukhatzhanova, *10th NPT Review Conference: Why It Was Doomed and How It Almost Succeeded Arm Control Today* (2022). https://www.armscontrol.org/act/2022-10/features/10th-npt-review-conference-why-it-was-doomed-and-how-it-almost-succeeded
13. R. Popp, L. Horovitz, A. Wenger (eds.), *Negotiating the Nuclear Non-Proliferation Treaty* (Routledge, Origins of the Nuclear Order. London, 2017)
14. W.C. Potter, The NPT & the sources of nuclear restraint. Daedalus **139**(1), 68–81 (2010)
15. J. Simpson, The 2000 NPT review conference, in *SIPRI Yearbook 2001, Armaments Disarmame2nt and International Security.* (Oxford University Press, Oxford, 2001), pp. 487–502
16. United Nations, *Treaty on the Non-Proliferation of Nuclear Weapons* (2024). https://treaties.unoda.org/t/npt
17. United Nations, *Nuclear-Weapon-Free Zones* (2024). https://disarmament.unoda.org/wmd/nuclear/nwfz/
18. United Nations Office of Disarmament Affairs, *Status of the Treaty on the Non-Proliferation of Nuclear Weapons* (2024). http://disarmament.un.org/treaties/t/npt

Chapter 15
The Obstacles to Restarting the Nuclear Arms Control and Disarmament Dialogue

Qizheng Zhang

Abstract This paper aims to address the important issue of how to restart nuclear arms control and disarmament dialogue. It analyzes the main obstacles to restarting current bilateral and multilateral nuclear arms control and disarmament dialogues, and concludes that the deterioration of bilateral political relations and different concerns about future arms control arrangements are the two main obstacles to restarting the U.S.-Russian bilateral nuclear dialogue. The main obstacles to restarting the multilateral nuclear dialogue are dissatisfaction among the non-nuclear weapon states with the pace at which the nuclear-weapon states effectively fulfill their nuclear disarmament obligations, and the urgent need of the non-nuclear weapon states for security assurances. In order to promote the restarting of the U.S.-Russian bilateral nuclear dialogue, the paper argues that efforts could be made to improve bilateral relations, respect each other's core interests, and jointly maintain strategic stability. To promote the restarting of multilateral nuclear dialogue, the nuclear-weapon states could utilize the P5 process to contribute to the international arms control process and respond collectively to the demands of the non-nuclear weapon states.

15.1 Introduction

For the past three decades, the Amaldi Conference has played a key role in reducing global nuclear risks and encouraging international cooperation on arms control [1]. However, we are living in a "problematic time", characterized by deteriorating relations among the major powers, intensifying geopolitical conflicts, a significant increase in nuclear risks, and an uncertain future for the international arms control process. Therefore, it is important to discuss the issue of restarting nuclear dialogue.

Q. Zhang (✉)
Center for Strategic Studies, China Academy of Engineering Physics, Beijing, China
e-mail: zhang_qizheng@iapcm.ac.cn

P. Cotta-Ramusino et al. (eds.), *Nuclear Weapon Risks in a Problematic Time*, Springer Proceedings in Physics 358, https://doi.org/10.1007/978-3-032-29873-7_15

15.2 The Stagnation of Nuclear Dialogue and Its Causes

At the level of bilateral nuclear disarmament and arms control, the United States and Russia possess the vast majority of the world's nuclear weapons and therefore have a special responsibility to prioritize disarmament. Consequently, U.S.-Russian bilateral nuclear disarmament and arms control forms the foundation of the global multilateral nuclear disarmament and arms control process. Historically, U.S.–Russian bilateral arms control has mitigated the destabilizing impact of the arms race on strategic stability and facilitated dialogue between the two parties. However, changes in the international security environment have led to setbacks in U.S.-Russian bilateral nuclear arms control. Following the expiration of the ABM and INF treaties, Russia announced a suspension of its participation in the New START Treaty in 2023, a move that was reciprocated by the United States with a suspension of on-site verification. Although research indicates that both parties are currently in compliance with the treaty's central limits, the treaty is due to expire in 2026. The absence of further arrangements will have a knock-on effect on the international arms control process. In the short term, it will be very difficult to forge new arms control arrangements within the remaining time frame. Therefore, promoting the resumption of nuclear dialogue is an important first step towards the next stage of arms control.

Currently, there are two main reasons preventing the United States and Russia from resuming nuclear dialogue.

Firstly, the deterioration of political relations directly impedes the resumption of nuclear dialogue. The deterioration of relations between the two countries affects the conduct of nuclear dialogue. This has happened before: the U.S.–Soviet rivalry during the Cold War led to the stagnation of their arms control process, and it was not until relations eased that nuclear dialogue resumed. The nuclear dialogue itself also served to ease tensions. Secondly, the two countries have different concerns about the future of arms control. Russia is concerned about the number of non-deployed U.S. nuclear weapons and their non-nuclear strategic technologies, such as missile defense. The United States, meanwhile, is concerned about the number of Russian non-strategic nuclear weapons. These differing concerns serve as an additional impediment to resuming the dialogue. This is also an obstacle to resuming cooperation.

In terms of multilateral arms control and disarmament dialogue, the most significant nuclear disarmament and arms control process currently underway is the NPT Review Conference process. However, this process has failed to produce results for the last two Review Conference cycles, much to the disappointment of the non-nuclear weapon states. The contradiction between the nuclear-weapon states and the non-nuclear weapon states on the issue of nuclear disarmament is deepening, and the nuclear dialogue is almost at a standstill.

There are two main reasons for this situation: firstly, non-nuclear weapon states feel that the nuclear-weapon states are not fulfilling their obligations under Article VI of the NPT Treaty effectively and that the pace of nuclear disarmament is too slow. They want to accelerate it, as evidenced by the TPNW treaty adopted by the non-nuclear states in 2017. In the current time of rising nuclear risks, they are more concerned and anxious.

Secondly, the security needs of non-nuclear weapon states have not been assured. During the NPT Review Conference cycle, the non-nuclear weapon states have emphasized the urgent need for unconditional negative security assurances. They have been frustrated by the persistent failure to meet their needs.

15.3 Recommendations to Facilitate the Restarting of Nuclear Dialogue

At the bilateral level, I believe that the United States and Russia could each take action in three areas.

Firstly, they could promote the improvement of political relations and create conditions for resuming dialogue. For example, they could actively ease tensions, resolve regional issues, and maintain regional security. Secondly, they could adopt a responsible and proactive attitude, respect each other's core interests, and take the initiative in demonstrating goodwill and rebuilding mutual trust. In the event of tense relations and a lack of mutual trust, any unilateral provocative behavior could lead to misunderstandings and miscalculations. Therefore, it is advisable for each side to exercise restraint and caution unilaterally, avoid further provoking the other side and refrain from testing or crossing the other side's "red lines" and "bottom lines".

Thirdly, they could respect their mutual vulnerability and work together to maintain strategic stability. Future dialogue on arms control between the two countries could consider not only limiting the number of nuclear weapons, but also emphasizing the maintenance of strategic stability. Future bilateral arms control arrangements could involve integrating strategic offensive and defensive capabilities, as well as deployed and non-deployed nuclear weapons. Respecting the survivability of each other's nuclear forces is also essential.

At the multilateral level, in order to restart effective dialogue with the non-nuclear weapon states, the nuclear-weapon states could explore efforts in three areas:

Firstly, they could collaborate with the non-nuclear weapon states to explain that achieving nuclear disarmament gradually and in an orderly manner is a more realistic and feasible way of ultimately resolving the nuclear disarmament issue.

Secondly, the nuclear-weapon states are encouraged to cooperate through the multilateral dialogue platforms and make substantive contributions to the international arms control process. The P5 process is a good platform, and the P5's joint statement in 2022 that "nuclear war cannot be won and must never be fought" is an important achievement [2].

In terms of the direction of cooperation, the P5 countries could discuss joint efforts to maintain strategic stability, reduce nuclear risks and study the impact of emerging technologies, such as AI, on nuclear arms control. Dialogue could also be initiated on arrangements to reduce the role of nuclear weapons, as well as on the mutual no-first-use of nuclear weapons. This would demonstrate a responsible attitude towards the non-nuclear weapon states.

Thirdly, the nuclear-weapon states could jointly respond to the demands of the non-nuclear weapon states. China has made a clear and unconditional commitment not to use or threaten to use nuclear weapons against non-nuclear-weapon states or nuclear-weapon-free zones, [3] and the issue of providing security assurances to the non-nuclear weapon states is an area for cooperation.

References

1. K. Gottstein, *The Amaldi Conferences. Their past and Their potential future* (2013). https://doi.org/10.48550/arXiv.1109.1572
2. Ministry of Foreign Affairs of the People's Republic of China, *Joint Statement of the Leaders of the Five Nuclear-Weapon States On Preventing Nuclear War and Avoiding Arms Races* (2022). https://www.fmprc.gov.cn/eng/xw/zyxw/202405/t20240530_11330619.html
3. People's Republic of China, *Implementation of the Treaty on the Non-Proliferation of Nuclear Weapons in the People's Republic of China* (2021). NPT/CONF.2020/41 https://documents.un.org/doc/undoc/gen/n21/341/55/pdf/n2134155.pdf

Part III

Chapter 16
Climate Change Following Nuclear War: A Brief Introduction

David Alexandre Ellwood

Abstract The five recognized nuclear weapons states have a legal obligation to pursue disarmament under Article VI of the NPT, but as the Doomsday Clock of the Bulletin of Atomic Scientists ticks ever closer to midnight, these same states are pursuing a costly modernization program that enshrines a nuclear oligarchy over our indefinite future. As we approach the 70th anniversary of the announcement of the Russell-Einstein Manifesto the scientific case to abandon nuclear deterrence has never been stronger. Modern climate models show that the global cooling induced by even a limited nuclear war would be devastating to the world community. This article provides a brief introduction to recent work on nuclear winter, and makes the case that nuclear weapons are not an instrument of national security, but rather a spectre of our own annihilation.

16.1 Speak Softly…and Carry a Big Stick

> Igitur quī̄ dēsī̄derat pācem, præparet bellum.
>
> (Flavius Vegetius Renatus)

Deterrence and diplomacy are supposed to work alongside one another, but do we have the balance right? How much weight should each be given? Or is the correct balance dynamic? Roughly speaking, we are talking about the "Stick" and "Carrot" approach to international relations. But does a big stick really confer greater security? That depends on who feels threatened by it, and more importantly, how they might react to that threat. Even if you believe in big sticks, the pertinent question is—how big is big enough? And big enough for what? A nuclear defence posture must answer such questions, and the answers need to address what sort of attack one intends to deter and by whom. But as soon as we adopt such an approach, how can we avoid falling into an endless cycle of military build up? A race to build ever larger and more lethal sticks?

D. A. Ellwood (✉)
British Pugwash, London, UK
e-mail: kangu@me.com

P. Cotta-Ramusino et al. (eds.), *Nuclear Weapon Risks in a Problematic Time*, Springer Proceedings in Physics 358, https://doi.org/10.1007/978-3-032-29873-7_16

If you want peace, prepare for war goes the old Roman adage, but if we adopt a policy of nuclear deterrence, what sort of war are we preparing for? Even defining what we mean by a nuclear deterrent poses difficulties, because the better we understand the nature of nuclear weapons, the harder it is to cast them in a rational light. It's quite unclear if they increase or decrease our national security. Indeed, history suggests that even the biggest sticks often fail to confer the security gains nations seek.

One thing is sure, nuclear weapons introduce a striking asymmetry into international relations, one that favours aggression and inequality, rather than greater cooperation and harmony. This asymmetry, and the absence of a common security architecture for the global community, makes it more likely that states will seek the "protection" of a nuclear deterrent. That both Sweden and Finland recently joined a nuclear alliance is a symptom of the state of our international system. Finland's conventional armed forces are extremely capable, but a Finnish analyst recently told me that the reason his country sought NATO membership was its "nuclear umbrella".

So if modern well equipped conventional forces are no longer considered a sufficient deterrent, what nuclear capability must one add to address the latent insecurity that remains? To know what is "just right" in such an extreme concept of deterrence it's necessary to explore the limits; what is "too big" and "too small" in the perilous world of weapons of mass destruction?

16.2 Ask a Dinosaur…

> In everything she took care to imitate the third choice of Goldilocks, just right.
>
> (Margaret Atwood)

In 1981, the physicist and Nobel Laureate Luis Alvarez together with his son Walter, suggested that the Cretaceous-Paleogene (K-Pg) extinction was caused by a massive asteroid hurtling into Gulf of Mexico [1]. It's estimated that 75% or more of all species on Earth vanished in a geological instant.

In the geologic record, the K-Pg extinction event is marked by a thin layer of sediment called the Fatkito boundary which can be found throughout the world in both marine and terrestrial rocks. The Alvarez' noticed that this boundary clay exhibited unusually high levels of iridium, which is more common in asteroids than the Earth's crust. In the 1990s a massive crater under the Yucatan Peninsula—called *Chicxulub* (cheek-shoo-LOOB) after a nearby village—was identified as the site of the alleged impact. Analysis of core samples from the crater have provided strong evidence that the K-Pg boundary clay represents debris from an asteroid impact about 65 million years ago. Recent supercomputer simulations [2, 5] allow us to imagine the details of the impact in extraordinary detail. The simulations show that the geophysical features of the Chicxulub crater correspond to an impact by an asteroid 17 km in diameter, travelling approximately 12 km s^{-1}. We can imagine it as follows:

On an otherwise ordinary day in the late Mesozoic, a massive asteroid approximately 17 km in diameter approached Earth from the north east at an angle of 60° from the horizontal. Simulations suggest the asteroid had a density of approximately 2630 kg m^{-3} and ploughed into the Gulf of Mexico at 72000 m hr^{-1} The impact created widespread tsunamis, sent a seismic shock wave around the planet and left an impact crater 300 km wide and 3 km deep. The vaporised rock from the impact was lofted into the upper atmosphere where it condensed into sand-grain-sized spheres. These were heated as they fell back towards the Earth and produced the effect of a *broiler in space* which killed the dinosaurs and ignited most of the plant life on the surface [11]. Within a few hours a fast moving dust cloud encircled the planet and climate active gases from vaporised rock rich in Sulphur together with vast amounts of black carbon from the burning hellscape were ejected into the atmosphere.[1] Sulphuric acid aerosols quickly adhered to the heavier silicate ejecta inducing acid rain storms on a continental scale, causing episodic acidification of both terrestrial and aquatic ecosystems.

It is fascinating to note that the energy that precipitated this epochal catastrophe was simply due to the speed of the asteroid. A velocity of just a few km s^{-1} endows an inert rock with the explosive yield of TNT! If you want to *blow something up* you don't need explosive materials, only enough relative *kinetic energy*. For almost twenty years the US government tried to develop a missile defence system on the idea of seeding the sky over the former USSR with thousands of *Brilliant Pebbles*. During its boost phase, an intercontinental ballistic missile (ICBM) travels about 6 to 7 km s^{-1}, more than enough kinetic energy to wrought its own destruction if small[2] autonomous[3] "pebbles" could be directed into its path. Brilliant Pebbles was part of Regan's "Strategic Defence Initiative" (SDI), popularly know as Star Wars. The project was cancelled in 1993, but I mention it here because it was widely considered the most successful missile defence idea to emerge from the SDI's research, and may soon resurface now that advances in AI offer potential to render such technology more versatile.

Let us return to the question of deterrence and compute the TNT equivalent yield of the Chicxulub impact. The simulations show that the detailed geophysical features of the Chicxulub crater correspond with an impact of an asteroid 17 km in diameter traveling 12 km s^{-1} with a density $\sim$ 2630 kg m^{-3}. At this speed each gram of asteroid would be endowed with a kinetic energy of

$$\begin{aligned} E_{g^{-1}} &= \frac{1}{2} \times (0.001) \times (12000)^2 \\ &= \frac{1}{2} \times 10^{-3} \times (144 \times 10^6) \\ &= 72 \times 10^3 \mathrm{J}. \end{aligned}$$

[1] The soot from these fires, together with the iridium and microspherules is found in the Fatkito boundary layer around the world.

[2] About the size of a watermelon.

[3] Heatseeking.

Now $12\,\text{km}\,\text{s}^{-1}$ is a relatively modest speed on astronomical scales,[4] yet sufficient to endow the Chicxulub impactor with an explosive yield approximatly 18 times that of TNT! From the volume and mass estimates;

$$\begin{aligned}
Volume_{Chic} &\approx \left(\frac{4}{3}\right) \times 3.142 \times (8.5 \times 10^3\text{m})^3 \\
&\approx 4.189 \times 614.125 \times 10^9\text{m}^3 \\
&\approx 2573 \times 10^9\text{m}^3, \\
Mass_{Chic} &\approx (2630\text{kg}\,\text{m}^{-3}) \times (2573 \times 10^9\text{m}^3) \\
&\approx 6.767 \times 10^{15}\text{kg},
\end{aligned}$$

the total energy yield of the Chicxulub impactor would be;

$$\begin{aligned}
Energy_{Chic} &\approx (6.767 \times 10^{15}\text{kg}) \times (72 \times 10^6\text{J}\,\text{kg}^{-1}) \\
&\approx \left(\frac{487 \times 10^{21}\text{J}}{4 \times 10^6\text{J}}\right) \approx (122 \times 10^{15}\text{kg}_{TNT}) \\
&\approx 122 \times 10^6\text{Mt}.
\end{aligned}$$

This is about 150,000 times the size of the entire US strategic arsenal! It would obviously be absurd to build a *deterrent* with anything like the explosive yield of the Chicxulub impactor. A threat to cause a mass extinction event is not a deterrent, just a sign of insanity. But how much headroom do we have to play with?

16.3 Divide et Impera

Dividing and conquering is not a sustainable strategy...

(Julius Caesar)

A nuclear deterrent is not a single massive weapon, but an arsenal of much smaller weapons whose energy may be spread out to maximise their net damage efficiency (bomb damage/bomb mass). Nuclear war planners are mainly concerned with blast effects, which scale to the two third power of the yield.[5] In Hiroshima the 5 psi (peak overpressure) contour extended 1.4 km from hypocentre, so

$$\text{Area}(15\,\text{kt}; \geq 5\,\text{psi}) \approx 6.2\,\text{km}^2.$$

[4] The average orbital speed of Earth is $\approx 30\,\text{km}\,\text{s}^{-1}$.

[5] The fractions of a bomb's yield emitted as thermal radiation, blast, and ionizing radiation are essentially constant for all yields, but a blast wave dissipates its energy throughout the medium it traverses, and thus decreases in intensity as the cube root of radius of an imaginary sphere centered on the explosion.

Scaling up, the 5 psi contour for an airburst of a 90 kt weapon encircles an area of:

$$\begin{aligned}\text{Area}(90\,\text{kt}; \geq 5\,\text{psi}) &\approx \left(\frac{90}{15}\right)^{\frac{2}{3}} \times 6.2\,\text{km}^2 \\ &\approx 20\text{km}^2.\end{aligned}$$

An MIRV'ed[6] Trident II missile can carry up to twelve[7] W76-1 warheads, each with a yield of approximately 90 kt. Thus the total yield of the payload is about 1 Mt and

$$\begin{aligned}\text{Area}(12 \times 90\,\text{kt}; \geq 5\,\text{psi}) &\approx 12 \times 20\,\text{km}^2 \\ &\approx 240\,\text{km}^2.\end{aligned}$$

However

$$\begin{aligned}\text{Area}(1\,\text{Mt}; \geq 5\,\text{psi}) &\approx \left(\frac{1000}{15}\right)^{\frac{2}{3}} \times 6.2\,\text{km}^2 \\ &\approx 100\,\text{km}^2.\end{aligned}$$

Most buildings inside the 5 psi will collapse, injuries are universal and fatalities will be widespread. So nuclear planners haven't opted for smaller yields for any humanitarian reasons ...you get *more than double* the blast damage for the same yield!

The composition and topography of modern urban environments introduce significant uncertainties, but a useful rule of thumb to estimate fatalities is to assume everyone inside the 5 psi contour would be killed, and nobody outside. Since the 5 psi contour of a single W88 (455 kt) warhead airblast[8] extends over 4 km from ground zero, it's easy to convince yourself that countervalue strategies[9] are as scary as they are intended to be. Nevertheless, even the staggering fatality numbers war planners estimate do not portend an apocalypse. But it wasn't blast effects that killed the dinosaurs! Even a blast the magnitude of Chicxulub doesn't cause a mass extinction event...So what does?

---

[6] Multiple independently targetable reentry vehicle.

[7] Treaty limited to eight.

[8] 384 of which are currently deployed by the US Navy on submarine launched ballistic missiles (SLBMs).

[9] Which include targeting the most densely populated urban environments.

16.4 Impact Winter

Oh, when will you ever learn?

(Pete Seeger)

The impact of Chicxulub was something like a mosquito hitting your windscreen. Its effect on your car is negligible, but it leaves a hell of a mess to clean up…The heat and thermal radiation released at Chicxulub ignited wildfires on a continental scale. The intense heat generated hurricane force winds which lofted tens of thousands of teragrams of black carbon/soot high into the atmosphere. Unable to photosynthesise, forest, land and ocean plants died; ultimately 75% of all life on earth either froze or starved to death! The K-Pg extinction event was severe, global, rapid, and selective, eliminating a vast number of species. The event appears to have affected all continents at the same time—fossil pollen records devastation of the plant communities in areas as far apart as New Mexico, Alaska, China, and New Zealand.

The key mechanism at work in such catastrophic devastation is that of a *firestorm*—a violent thunderstorm involving both precipitation and extreme winds. As the hot air above large fires (or conflagrations) rises up it pulls in air from all directions. When the fire is particularly ferocious, these columns coalesce into a large convective column which is the hallmark of a *firestorm*—a weather system generated by the fire itself. As the central column of superheated combustion products continue to rise, radial winds take hold which pull in more dense cold oxygen-rich air. Tornado like vortices form, funnelling up massive quantities of soot which generate clouds that tower into the sky like vast smoke stacks. These clouds—called *pyrocumulonimbi*—eject soot into the upper troposphere. The soot is self lofting due to solar heating, and continues to rise up into the stratosphere. New research shows the soot lofted in this way rises to much higher altitudes than expected and is likely to remain five times longer than suggested by early (1980s) modelling.

Of course you don't need an asteroid impact like Chicxulub to generate a firestorm, just sufficient quantities of thermal energy and combustible materials. Nuclear explosions, wild fires and bombing raids can all ignite firestorms, but only if the conditions are right. London and Coventry were fire-bombed during World War II, but the raids didn't result in a firestorm because they were not sufficiently concentrated in both space and time. British war planners understood this and the raids on Hamburg (27 July 1943) and Dresden (13–15 February 1945) resulted in firestorms of 12 km^2 and 21 km^2 respectively. Based on WWII data Glasstone and Dolan [4] identified minimum requirements for a bombing raid to generate a fire storm:

1. at least 8 pounds of combustibles per square foot of fire area (40 kg per square meter);
2. at least half of the structures in the area on fire simultaneously;
3. a wind of less than 8 mi per hour at the time; and
4. a minimum burning area of about half a square mile.

The 15 kt atomic bomb dropped on Hiroshima (6 August 1945) generated a firestorm of 11 km^2, but the 21 kt bomb on Nagasaki didn't. One reason for this was different topography. Hiroshima is relatively flat while Nagasaki has an uneven landscape. Nevertheless, some 40,000 men, women and children died in Nagasaki. 70,000 more were injured, and radiation would continue to take its toll indefinitely. What is often omitted is that only 150 of the fatalities were members of the Japanese military. With or without a firestorm, the experience of Hiroshima and Nagasaki give new meaning to the term *collateral damage*. As the historian Margaret MacMillan remarked [6],

> It was no oversight that mass bombings were not included in the Allied indictment of Nazi leaders at the Nuremberg trials.

Firestorms may be[10] less likely in modern cities because of their more resistant construction and design. In general, city fire damage is less dependent on the yield of the weapons used than conditions in and around the city. Fuel loading per square meter (fuel density) is arguably the most significant factor. But if the weapons are powerful enough, a city will be completely destroyed by heat and blast effects alone, not to mention the irradiated rubble left in its wake. In summary, nuclear weapons reduce the question of whether a city will firestorm to a few predictable variables, topography, fuel loading and the prevailing atmospheric conditions (e.g., wind speed). With cities firmly in the crosshairs of nuclear war planners, large quantities of soot are certain to result if nuclear weapons are ever used. Just how large is our next subject…

16.5 Sootification

> The less you rely on smoke and mirrors, the less you'll need to put out fires.
>
> (Gavin Mills)

We can approximate [10, 13] the amount of fuel (combustible loading) per unit area (urban/developed landscape) by a linear function of population density:

$$M_f = (1.1 \times 10^4\,\mathrm{kg\,person}^{-1}).P + 8 \times 10^6\,\mathrm{kg\,km}^{-2}.$$

Estimates of population density are readily available, e.g., the *LandScan*[11] global population database provides daily average population in grid cells of $30'' \times 30''$ arc seconds (i.e., $\sim 1\,\mathrm{km}^2$ at the equator).

---

[10] A recent NAS report [7] suggests that firestorms are less likely under many plausible conditions than earlier models assumed but this conclusion is far from evident. Significant uncertainties abound, including: How much urban fuel is available and its combustibility (e.g. modern building materials vs older ones)? How do blast and thermal effects interact (thermal pulses igniting fires, blast waves causing secondary ignitions or creating rubble)? Are fires in densely built urban areas less likely to coalesce into large firestorms with strong convective columns? While urban environments populated with highrise buildings don't lend themselves to the formation of firestorms because of the baffle effect of these structures, a nuclear weapon will incinerate everything within a certain radius and the contents of a building may provide more fuel than its construction. This is an area of active research and new modelling is expected to shed more light on these matters soon.

[11] https://landscan.ornl.gov.

The total single detonation mass of soot emitted by fires (adjusted for rain out) is estimated [13] by the formula[12]:

$$M_{Soot} = \sum_{j=1}^{J} A_j M_{f_j} . \left(\sum_{i}^{N} F_i Q_i S_i C_i R_i \right) .$$

The first sum is over all grid cells subject to fire ignition, A_j is the area of the jth grid cell and M_{f_j} is the quantity of fuel per unit area of the jth cell estimated according to population density (by the formula above). The second sum is a factor which corrects for the different types of combustible materials present and how much soot they generate. In reality this would vary between grid cells but no attempt is made to take account of this. Recent work [10, 12] assumes all the fuel of each type is consumed ($Q_i = 1$), the fuel type per person in the city bombed is the same as it would be in the developed world ($C_i = 1$) and that soot rainout due to fire storms is 20% ($R_i = 0.8$). Summing over fuel types then gives a value of $0.016\,\text{kg}_{\text{Soot}}\,\text{kg}_{\text{Fuel}}^{-1}$ for the second sum. Substituting in the formula for population density then yields:

$$\begin{aligned} M_{Soot} &= \sum_{j=1}^{J} A_j M_{f_j} . \left(\sum_{i}^{N} F_i Q_i S_i C_i R_i \right) \\ &= \sum_{j=1}^{J} A_j \left(1.1 \times 10^4).P + 8 \times 10^6\right) .0.016 \\ &\approx \sum_{j=1}^{J} A_j \left(180\text{kg}_{Soot}\,\text{psn}^{-1}).P\text{psn}\,\text{km}^{-2} + 130,000\text{kg}_{Soot}\,\text{km}^{-2}\right) . \end{aligned}$$

At Hiroshima 13 km^2 burned. Linear scaling is sufficient[13] to calculate the area burnt for small yield weapons, e.g., the airblast of a 100 kt weapon burns:

$$\text{Area}_{Fire\ zone}(100\,\text{kt}) = 13\,\text{km}^2 \times \frac{100}{15} = 86.6\,\text{km}^2.$$

To understand the climate effects of a nuclear war, we must consider targeting. Deterrence theory favours a *countervalue* strategy in which urban areas and critical infrastructure are included amongst legitimate targets, rather than a focus on opposing nuclear forces (i.e., *counterforce* targeting). Thus the areas of highest population densities would be likely casualties of a nuclear war, and we can expect such a conflict to generate the maximum amount of soot suggested by the above formula. Civilian studies don't have access to the actual targeting maps of nuclear war planners, so

---

[12] This is the best we can do at the moment. However ongoing research is now modeling the burning of cities in detail and how much smoke they would produce.

[13] The area within a given thermal energy flux contour varies linearly with yield for small yield weapons.

recent studies adopt a simplistic *cookie cutter* approach to targeting, where warheads are directed at the areas of highest population density in a non-overlapping grid.

This is certainly contrary to actual war plans, in which huge redundancy is built in to ensure the desired blast damage. Further work with more accurate targeting plans will more accurately assess probable soot mass, but the effects are likely to be of the same order of magnitude because the distribution of combustible loading within possible adversaries is quite concentrated.

In their papers [8, 9, 12] Robock and his collaborators focus on two scenarios, a full scale nuclear war (United States/Russia), and a regional nuclear war (India/Pakistan). The full scale war scenario assumes that Russia targets 1000 × 100 kt-weapons on the US and 200 × 100 kt-warheads each on France, Germany, India, Japan, Pakistan, and the UK; while the US targets 1100 × 100 kt-weapons each on China and Russia[14]. They estimate as much as 180 Tg of soot could be generated in such a war. Such an exchange is a gross simplification that does not accurately depict the makeup of current arsenals[15], but more probable weapons yields would likely produce comparable soot emissions.

The regional war scenario assumes each side targets 50 × 15 kt weapons on the other, i.e., 100 nuclear explosions and 1.5 Mt total yield. For this exchange they estimate 6.6 Tg of soot may be generated. To understand the climate response in each case, these soot estimates are rounded down to 150 Tg and 5 Tg respectively and fed into supercomputer simulations called *Earth system models.*

16.6 The Science (and Art) of Climate Modelling

> All models are wrong, but some are useful.[16]
>
> (George Box)

The laws of physics (thermodynamics, conservation of mass/energy, etc.) that determine how energy and matter interact in the oceans, land and atmosphere are well known, but difficult to compute. The most sophisticated models of Earth's systems work by dividing the surface of our planet up into a fixed grid. A 3-D array of cells above and below the grid, reaching up into the atmosphere and down into the oceans, is used to build a numerical model of atmospheric and oceanic dynamics. These

[14] i.e., 4400 total nuclear explosions and 440 Mt total yield.

[15] 4400 explosions exceeds the capacity of current arsenals.

[16] This is a negative way of looking at it, but it drives home an important point to a lay audience. A leading practicioner suggested to me it's better to say "all models are incomplete and imperfect, but they are nevertheless very useful." For example, we use models every day to make weather forecasts. And of course, nuclear weapons research for the last several decades has relied solely on models in the absence of underground testing. Since the entire Earth is our laboratory, we must use simulations to ask "what if" questions without messing it up. But modelling complex systems is very challenging, and climate modelling pushes the boundaries in many ways.

models simulate the movement of energy, water, gases, aerosols etc. into and out of the 3-D array of grid cells. To achieve this *general circulation models* (GCMs) as they are called, employ a supercomputer to integrate discrete versions of the *Navier-Stokes equations* on a rotating sphere together with thermodynamic terms for various energy sources (radiation, latent heat, etc.). Separate models for the Ocean (OGCM) and atmosphere (AGCM) are coupled together with submodels[17] so as to simulate Earth's systems as comprehensively as possible.

The simplest grid used in AGCMs consists of a constant angular grid (i.e., a latitude/longitude grid) with resolutions between 1° and 5°. One degree of both latitude and longitude at the equator covers approximately 111 km.[18] This distance is consistent for latitude because lines of latitude are parallel. However, for longitude this distance applies only at the equator—the distance between longitude lines decrease as they converge at the poles.[19] In any case, current models are too coarse to capture important small-scale features such as clouds, and modellers must make *subgrid assumptions* to account for this. I think it's fair to say that there is both science and art involved in climate modelling. Even the best models may fail to reproduce past climate events and some *fine tuning* is often required. This means adjusting known parameters in such a way that the model behaves plausibly enough to be useful for the various climate scenarios under investigation.

Climate modelling is a very active area of research, and even with improvements in computer power, different models may disagree when attempting to simulate subtle phenomena, including anthropogenic warming. However, there is nothing subtle about the severe shock to Earth's systems caused by a nuclear war, even in the regional war scenario. In particular, AGCM predictions of *nuclear winter* exhibit the same narrow range of results, and these results are very alarming. If you feel confident in basing energy policy on the modelling results, you cannot ignore the strong message these models send about the climate risks of a security posture that includes nuclear deterrence.

16.7 Up in Smoke...

> Models made by Russian and American scientists showed that a nuclear war would result in a nuclear winter that would be extremely destructive to all life on Earth; the knowledge of that was a great stimulus to us, to people of honor and morality, to act in that situation .
>
> (Mikhail Gorbachev)

---

[17] To account for evapotranspiration, sea-ice, chemical and other processes.

[18] The Earth is approximately a sphere, and its circumference is about 40, 000 km. We divide a circle into 360°. Therefore, each degree of a great circle on earth's surface is roughly 40, 000 km/360 = 111.111 km in length (60 nautical miles). While the Earth is not a perfect sphere (it is an oblate spheroid), this number provides a practical approximation.

[19] Hence AGCM's must impose constraints on variables assigned to cells close to the poles so as to prevent instabilities developing.

In their 2007 work, Robock and his collaborators [8, 9] employ the NASA Goddard Institute for Space Studies (GISS) ModelE to explore the climate response in both the 150 Tg and 5 Tg scenarios. The atmospheric portion of the model was run at $4^\circ \times 5^\circ$ latitude-longitude resolution with 23 vertical layers extending to 80 km. The coupled oceanic general circulation model had 13 layers and also a $4^\circ \times 5^\circ$ latitude-longitude resolution. The GISS ModelE includes a module to calculate the transport and removal of aerosol particles and is reported to have performed well in simulating the climatic response to both the 1783 Laki and the 1912 Katmai volcanic eruptions. It has also been used to successfully simulate the transport and removal of sulfate aerosols from tropical and high-latitude volcanic eruptions.

In the full scale nuclear war scenario, 150 Tg of black carbon was injected into the upper troposphere over 7 d starting on 15 May, equally spread over cells covering the contiguous United States and Russia. In the 5 Tg scenario all the black carbon was injected on a single day (May 15), and concentrated into a single column of grid boxes at 30° N, 70° E. There were some notable differences in the way smoke particles were represented in each case [9]. In the latter, the optical properties of black carbon aerosols were set to those of pure smoke to more accurately reflect burning from city targets, as opposed to a general mixture of black soot, smoke, and dust that was modeled in the the 150 Tg case. Thus per unit mass, the aerosols in the 5 Tg scenario would be expected to absorb more solar radiation and producing more lofting, and hence induce a more extended perturbation of the climate following a nuclear war.

Climate modelling is not weather forecasting, where a great deal of the uncertainty results from the chaotic nature of the dynamical system, i.e., its sensitivity to the specification of initial conditions. Rather, climate modelling looks for effects that are independent of such small variations, and so an *ensemble* of trials is considered based on a suitably random distribution of initial conditions. Robock and his collaborators conducted three ten year runs with 5 Tg of soot and compared the mean with a thirty year control run. The variation between ensemble members was small compared to the response, and a single run was considered sufficient for the very large forcing following from a 150 Tg soot injection.

Of course it's good practice not to rely on the predictions of computer simulations blindly, but to perform "back of the envelope" calculations to get a feeling for the relevant physics at play..

16.8 Nuclear Forcing

> The prediction of nuclear winter is drawn not, of course, from any direct experience with the consequences of global nuclear war, but rather from an investigation of the governing physics.
>
> (Carl Sagan)

The Sun is the power house of the Earth's systems. The solar constant $\Phi_S \approx 1361\,\mathrm{W\,m^{-2}}$ indicates the average power of the Sun's rays incident on our planet. A maximum of about $1000\,\mathrm{W\,m^{-2}}$ reaches sea level on a clear day at noon.[20]

Now it's a matter of common experience that hot objects glow, and the warmer they are, the brighter they glow. In fact all objects glow, but the colours of light/radiation they emit are often too faint to see, or outside of our visible range. It turns out that any object emits dramatically more radiation when it is hot than when it is cool, the exact relation being given by the famous *Stefan-Boltzmann law*:

$$\Phi = \sigma T^4. \quad \text{(SB)}$$

Here Φ is the total[21] *radiation flux*[22] emitted by a *blackbody* with *absolute temperature* T, i.e., an idealized object that absorbs and emits radiation of all wavelengths with 100% efficiency, $\sigma = 5.68 \times 10^{-8}\,\mathrm{W\,m^{-2}\,K^{-4}}$ is the Stefan-Boltzmann constant and T is *absolute temperature* which is measured in degrees Kelvin (K).[23]

The Earth is not a blackbody at visible wavelengths,[24] but radiates almost exclusively in the infrared band.[25] Hence we can approximate the terrestrial emission flux as that of a blackbody with global surface mean temperature T_E. Several complications preclude us from estimating T_E directly from the ambient flux, but we can rearrange the Stefan-Boltzmann law to derive a rough and ready estimate for how *changes* in the Earth's energy budget alter T_E. Rearranging (SB) for T and (Taylor) expanding:

$$T_E(\Phi + \Delta\Phi) = T_E(\Phi) + T_E'(\Phi).\Delta\Phi + \cdots$$

where $T_E'(\Phi)$ is the *derivative*;

$$T_E'(\Phi) = \frac{1}{4}\left(\frac{T_E(\Phi)}{\Phi}\right),$$

gives

$$T_E(\Phi + \Delta\Phi) = T_E(\Phi) + \frac{1}{4}\left(\frac{T_E(\Phi)}{\Phi}\right).\Delta\Phi + \cdots$$

and so

---

[20] This is a measure of the energy of sunlight arriving each second on a horizontal surface of one square meter normal to the Sun's rays, i.e., the maximum solar radiation flux at sea level.

[21] All wavelengths/frequencies.

[22] Energy per unit time per unit area.

[23] $^\circ\mathrm{K} = {}^\circ\mathrm{C} + 273.15$..

[24] The absorption efficiency of solar radiation by the Earth is only $\epsilon = 1 - \alpha = 0.72$, where α is called the *planetary albedo*.

[25] i.e., wavelenghts longer than visible light but shorter than the microwaves used by your cell phone.

$$\frac{\Delta T_E}{T_E} = \frac{T_E(\Phi + \Delta\Phi) - T_E(\Phi)}{T_E(\Phi)} \approx \frac{1}{4}\left(\frac{\Delta\Phi}{\Phi}\right) \qquad \Delta\text{SB}$$

i.e., the percentage change in temperature will be approximately one fourth the percentage change in the global average downward energy from the Sun reaching the Earth's surface.

The simulations performed in [8] suggest a 5 Tg soot injection would reduce the global average downward energy from the Sun reaching the surface by as much as $16\,\text{W m}^{-2}$ in the first year. Substituting $\Delta\Phi = -16\,\text{W m}^{-2}$ into (ΔSB) suggests an associated percentage change in temperature of about -0.4%. If we take T_E to be $290\,^\circ$K we find $\Delta T_E \approx -1.2\,^\circ$C, which is consistent with computer modelling predictions.

For comparison, the 1991 Mt. Pinatubo eruption – the largest of the twentieth century – reduced global average sunlight by only $4\,\text{W m}^{-2}$ with an observed temperature anomaly of about $-0.5\,^\circ$C. However, compared with forcing from volcanic ash, the forcing following a nuclear war would persist for years. Thus even a nuclear war involving as little as only 0.03% the explosive power of the current world arsenal could induce climate change unprecedented in the last 1000 years.

Similarly, modelling shows that a 150 Tg soot injection may reduce the global average downward energy from the Sun reaching the surface by as much as $100\,\text{W m}^{-2}$ in the first year. Substituting $\Delta\Phi = -100\,\text{W m}^{-2}$ into (ΔSB) suggests an associated percentage change in temperature of about -2.5%, i.e., a temperature anomaly $\Delta T_E \approx -7.25\,^\circ$C. The simulation [9] predicts a comparable drop in the first year, but an even more extreme anomaly of $-9\,^\circ$C follows in the second year. This demonstrates the power of feedback effects[26] not captured by our simple energy balance estimates. Computer modelling also shows that recovery may be very slow, with an anomaly of $-4\,^\circ$C remaining one decade after the soot injection.

In 2019, Coupe, Bardeen, Robock and Toon [3] repeated the 150 Tg injection scenario with the Whole Atmosphere Community Climate Model version 4 (WACCM4), run for twenty years at $1.9^\circ \times 2.5^\circ$ resolution with 66 vertical layers.[27] This run incorporated full stratospheric chemistry and more accurately modeled the evolution of the black carbon aerosols, allowing for particle growth. They found global average downward energy from the Sun would be reduced by $120\,\text{W m}^{-2}$ in the first year, even more extreme than that predicted in their 2007 study. However, the WACCM4 run suggested a faster recovery, since the soot lifetime is reduced. Nevertheless, in this model the global climate fails to recover completely, even after twenty years.

The possibility of such massive forcing of the earth's climate following nuclear war poses serious questions for policy makers. More radiative heating of the stratosphere results in ozone depletion. Less radiative heating of the oceans means less precipitation. Also, lower temperatures globally result in shorter growing seasons, and the simulations suggest the possibility of freezing temperatures occurring during

[26] e.g., Lower temperatures result in more snow and ice (increased albedo) and less atmospheric water vapour (reduced greenhouse effect).

[27] up to 140 km.

summer months for years following a nuclear war. Ultimately a nuclear war would result in a global famine [15], a conclusion that fundamentally undermines the idea of a nuclear deterrent. How does a defence posture that implies national suicide increase the security of any nation?

So if current arsenals are too large, perhaps more nations should adopt a policy of *minimal* deterrence to allay these concerns?

16.9 Is One SSBN One SSBN Too Many?

> Every thinking person fears nuclear war and every technological nation plans for it. Everyone knows it's madness, and every country has an excuse.
>
> (Carl Sagan)

The United Kingdom has been a nuclear power since 1952. It is one of the five officially recognised nuclear states under the non-proliferation treaty (NPT). At its Cold War peak the UK nuclear stockpile consisted of approximately 520 nuclear warheads. Following the end of the Cold War, a review of the UK's nuclear posture resulted in the phase out of the the RAF's WE-177 free-fall bombs. The capability of the Royal Navy's surface ships to carry or deploy nuclear weapons was also dismantled and by 1998 the UK's deterrent was reduced to a single system.

The UK adopts a posture of *minimal credible nuclear deterrence* assigned to the defence of NATO. It is the only nuclear weapon state that has reduced to a single deterrent system. That system consists of:

- Four Vanguard-class submarines (SSBN), maintaining continuous at sea deterrence (CASD);
- 58 Trident II D5 missiles which are held in a communal pool with the US, situated at the Kings Bay Submarine Base in Georgia;
- Varying numbers of Mk4/A nuclear warheads that can be deployed on the Trident II D5s.

In 2010 the UK Government published official information on the size of the UK's overall nuclear stockpile for the first time (confirming that the UK held 225 warheads). However, in the 2021 Integrated Review the cap on the UK's stockpile was raised to (no more than) 260 warheads. Nevertheless, the UK retains the smallest inventory of any recognised nuclear weapon state.

However, the fact remains that only a small number of weapons are needed to destroy the cities and economies of even the largest countries. Indeed, maintaining a *minimal* deterrent *means* retaining sufficient force to destroy at least 25% of the population and 50% of the industry of any possible adversary! This concept is problematic at best:

- It's contrary to international humanitarian law (a minimal deterrent relies on a countervalue strategy, but threats or use of force must not be directed at civilians);
- Its impacts are disproportionate and also affect areas remote from the war zone;
- Security guarantees are not credible (whatever a minimal deterrent is, it's not an "umbrella");
- It ignores the threat felt by others, which increases insecurity!

But climate modelling suggests there is worse![28]

The concept of *minimal* deterrence is based on retaining the capability to inflict blast damage of sufficient scale to deter any possible adversary. But as we have seen, the destructive force of nuclear weapons is not limited to blast effects. Although less predicable, the effects of fire cannot be disregarded, and modern climate models suggest any nuclear war could result in a very bleak outcome for the entire planet [15]. Assuming linear scaling the airblast of a 90 kt W76-1 warhead could incinerate:

$$\text{Area}_{\textit{Fire zone}}(90\,\text{kt}) = 13\,\text{km}^2 \times \frac{90}{15} = 78\,\text{km}^2.$$

An MIRV'ed Trident II missile can carry up to twelve[29] such warheads. Hence

$$\begin{aligned}\text{Area}_{\textit{Fire zone}}(12 \times 90\,\text{kt}) &\approx 12 \times 78\,\text{km}^2\\ &\approx 936\,\text{km}^2.\end{aligned}$$

A fire zone of this scale can generate over 3.5 Tg of soot depending on combustible loading. Each Vanguard-class submarine (SSBN) can carry 8 MIRV'ed Trident II missiles, and hence up to 96 W76-1 warheads, but currently UK policy allows for a maximum of just 40 such warheads:

$$\begin{aligned}\text{Area}_{\textit{Fire}}(40 \times 90\,\text{kt}) &\approx 40 \times 78\,\text{km}^2\\ &\approx 3120\,\text{km}^2.\end{aligned}$$

Who and what this weapon is intended to deter is left open. The UK has a policy of deliberate ambiguity and does not make clear the precise details of when, how and at what scale the UK may consider the use of its nuclear weapons. However both the 2021 Integrated Review of Security, defence, Development and Foreign Policy and its subsequent refresh in March 2023 confirmed that nuclear weapons would only be used in "extreme circumstances of self-defence" and that the UK "will not use, or threaten to use, nuclear weapons against any non-nuclear weapon state party to the NPT".

Combustible loading of Russian cities is lower than India/Pakistan, but combustible loading of Chinese cities is comparable with the latter. Thus the *deterrent* provided by a single Vanguard-class submarine carrying 40 W76-1 warheads risks

[28] c.f., [14].

[29] Currently limited to eight.

climate forcing associated with a soot injection between 7 Tg and 12 Tg. This correlates to a crash in global surface temperatures of between −1.6 °C to −2.8 °C. Thus the only thing *credible* about a *minimal* deterrent is that more than a billion people with marginal food supplies could die of starvation. Isn't it time we rethought the concept of nuclear deterrence before current modernisation programmes condemn us to this absurd state of affairs for our indefinite future? As Andrew Carnegie said at the 1st National Arbitration and Peace Congress more than a century ago:

> These vast armaments on land and water are being defended as a means, not to wage war, but to prevent war (...) there is a safer way (...) it requires only the consent and the good-will of the governments. Today they say (...) If you want peace, prepare for war. This Congress says on behalf of the people: *Si vis pacem, para pactum*, if you want peace, agree to keep the peace.
>
> (Andrew Carnegie)

Acknowledgements I would like to thank *British Pugwash* for support and encouraging me to speak on this topic. I would also like to thank the *Physicists Coalition for Nuclear Threat Reduction* for introducing me to recent work on nuclear winter and Professor Alan Robock for numerous helpful comments and suggestions.

References

1. L.W. Alvarez et al., Extraterrestrial cause for the cretaceous-tertiary extinction. Science **208**(4448), 1095–1108 (1980)
2. G.S. Collins et al., A steeply-inclined trajectory for the Chicxulub impact. Nat. Commun. **11**(1), 1480 (2020)
3. J. Coupe et al., Nuclear winter responses to nuclear war between the United States and Russia in the whole atmosphere community climate model version 4 and the Goddard institute for space studies ModelE. J. Geophys. Res.: Atmos. **124**(15), 8522–8543 (2019)
4. S. Glasstone et al., *The Effects of Nuclear Weapons* (U.S, Department of Defense, 1977)
5. S.P.S. Gulick et al., The first day of the Cenozoic. Proc. Natl. Acad. Sci. **116**(39), 19342–19351 (2019)
6. M. MacMillan, *War: How Conflict Shaped Us*. (Random House Publishing Group, 2020), ISBN: 9781984856142
7. Engineering National Academies of Sciences and Medicine, *Potential Environmental Effects of Nuclear War*. (The National Academies Press, Washington, DC, 2025), ISBN: 978-0-309-71617-8
8. A. Robock et al., Climatic consequences of regional nuclear conflicts. Atmos. Chem. Phys. **7**(8), 2003–2012 (2007)
9. A. Robock, L. Oman, G.L. Stenchikov, Nuclear winter revisited with a modern climate model and current nuclear arsenals: still catastrophic consequences. J. Geophys. Res.: Atmos. **112**(D13) (2007)
10. O.B. Toon et al., Atmospheric effects and societal consequences of regional scale nuclear conflicts and acts of individual nuclear terrorism. Atmos. Chem. Phys. **7**(8), 1973–2002 (2007)
11. O.B. Toon et al., *Earth in Flames: How an Asteroid Killed the Dinosaurs and How We Can Avoid a Similar Fate from Nuclear Winter*. (Oxford University Press, 2025), ISBN: 9780197799703
12. O.B. Toon, A. Robock, R.P. Turco, Environmental consequences of nuclear war. Phys. Today **61**(12), 37–42 (2008)

13. R.P. Turco et al., Climate and smoke: an appraisal of nuclear winter. Science **247**(4939), 166–176 (1990)
14. P. Webber, Could one Trident submarine cause 'nuclear winter'? SGR Newsletter **124**(35) (2008)
15. L. Xia et al., Global food insecurity and famine from reduced crop, marine fishery and livestock production due to climate disruption from nuclear war soot injection. Nature Food **3**(8), 586–596 (2022)

Chapter 17
International Science Cooperation and Science Diplomacy: Challenges and Opportunities

Persis S. Drell

Abstract In this article I will begin by clarifying how I think about international science cooperation and science diplomacy and give a few examples of the roles they have played in the past. I will then talk about what has changed and what some of the current challenges are. I'll finish with a few thoughts on a path forward to ensure that we can continue to have the benefits of international science cooperation and science diplomacy in a changing world. I hope to convey two messages in this talk. The first is that I believe that international science collaboration and science diplomacy are essential tools to address global challenges and to support work towards global equitability and stability. Great science and tremendous progress towards international security have been supported in past decades by science collaboration and diplomacy. I hope this can continue in the future. My second message is that successful collaboration and diplomacy rely on an open international research community. I see threats to this open community, and I encourage scientists to continue to invest time and energy to preserve it. I want to acknowledge that my viewpoint in this paper will undeniably be US centric.

17.1 Why Is International Science Cooperation Important?

In my judgement the single most powerful argument for international science cooperation and collaboration is that it will result in better science. As the community broadens internationally, we are exposed to new ideas, and we have our own ideas tested, extended and validated.

However, there are additional powerful arguments that are worth mentioning. In many areas of science, the scale of the science requires international collaboration and cooperation. In fields such as accelerator-based particle physics, fusion, and space exploration, the tools of discovery are so expensive and complex that no one nation or region has the resources or skills to build and operate them. Additionally,

P. S. Drell (✉)
Stanford University, Stanford, CA, USA
e-mail: persis@stanford.edu

P. Cotta-Ramusino et al. (eds.), *Nuclear Weapon Risks in a Problematic Time*, Springer Proceedings in Physics 358, https://doi.org/10.1007/978-3-032-29873-7_17

there are fields such as climate science where global engagement is needed to have access to the data for developing and testing reliable models.

Global science cooperation allows us to become aware of the capabilities of our collaborators and allows us to meet the best talent world-wide. And finally, it is in our national interest. We can project our values of ethics and transparency, we can enhance our security based on demonstrated intellectual knowledge and capability, and it allows us to establish channels of communication that can overcome technical or political hurdles, and lead to broader international discussions in science and beyond.[1]

17.2 What Is Science Diplomacy?

There are at least three concepts that are referred to by the phrase 'science diplomacy' and they are well described in many references, a few of which are listed here [1, 2].

Reviewing those concepts briefly we have.

17.2.1 Diplomacy for Science

Diplomacy can facilitate the creation of international science facilities that are too large and too complex to be constructed and operated by a single nation. One example is the international treaty organization, CERN, that is exploring the frontiers of particle physics with powerful accelerators. Another example is ITER, the world's largest fusion experiment.

17.2.2 Science for Diplomacy

Science for Diplomacy describes efforts to leverage science engagement and exchange to advance broader national interests beyond science discovery [3]. The 2024 Biennial Report to Congress on International Science & Technology Cooperation states "International S&T cooperation has been a pillar of US foreign policy and national security since the end of WWII".[2] Science for diplomacy is especially important when political differences or economic competition are straining normal

---

[1] Material in this section benefitted from an unpublished document from and discussions with Professor Raymond Jeanloz of UC Berkeley

[2] NSTC Biennial Report to Congress on International Science & Technology Cooperation, February 2024. https://bidenwhitehouse.archives.gov/wp-content/uploads/2024/02/2024-Biennial-Report-to-Congress-on-International-Science-Technology-Cooperation.pdf

diplomatic relations. The continued interactions of many physicists with their Soviet counterparts during the cold war is an example of this.

17.2.3 Science in Diplomacy

Many international challenges, such as climate change, pandemic management, and verification of nuclear treaties, are increasingly scientific in nature. Science based policies are vital for success in these areas, and engagement of the scientific community with the policy makers is essential. The work of Tom Neff who proposed the conversion of highly enriched uranium from old weapons in the Soviet Union into fuel for nuclear power plants in the United States at the end of the cold war is an excellent example. The plan reduced the global stockpile of nuclear weapons, removed fissile material from the market and helped stabilize the world post 1991 [4].

17.3 What Has Changed and What Are the Threats to the Open Science Ecosystem Today?

The world is a very different place in 2025 than it was in 1990 or 2000. On the one hand, science is more internationally collaborative than ever before. On the other hand, the open global collaborative research environment in the US is facing new challenges.

The development of the internet over the past 2–3 decades has democratized cutting-edge scientific research and knowledge and has made the world of science very 'flat'. As soon as a researcher is ready to share a new result, it can be accessed by colleagues around the globe almost instantly. Science is more accessible in more places than ever before, facilitating rapid innovation globally. Collaboration is easier than ever, and the international engagement on major scientific projects now includes many more countries and nationalities than in previous decades. As information travels literally at the speed of light around the globe, a very real danger is being left behind and getting surprised by new discoveries if a nation chooses not to engage fully in the global scientific ecosystem. But there are multiple emerging threats to our open science ecosystem.

China is the most prolific international collaborator of the United States.[3] However, for some in the US, the rise of China as an economic power is seen as a threat to US leadership in science and technology. As a result, China is being viewed by many in the US and elsewhere as more of a threat than a collaborator. This is exacerbated by increasing concern with inappropriate behavior on the part

[3] The latest available published information on international collaboration by the NSF includes data up through 2018. NSF Science and Engineering Indicator, December, 2019, https://ncses.nsf.gov/pubs/nsb20206/international-collaboration

of the Chinese government to acquire intellectual property and critical emerging technologies for economic gain among other goals.[4]

Many powerful new technologies that are currently being developed have dual use potential. There are national security concerns with emerging technologies such as AI, QIS and Biotech. These concerns are fueled by observations that technologies seem to be moving up technical readiness levels very quickly from basic research to proof of concept and prototype. When and how should these new areas be protected?

Economic and national security concerns are leading some to advocate for restricting access to emerging new technologies ever earlier in the research process. This is complicated in the US by a blurring of civilian and military objectives.

Another threat to the open science ecosystem in the US comes from changes in the US research university system. There has been a shift in universities in recent decades to focus on entrepreneurship and generation of technology and innovation. At least partially, this arises from the landmark Bayh-Dole legislation[5] that, in 1980, enabled universities to own, patent and commercialize inventions derived from federally funded research. As a result, we are seeing more technology development in our universities and a resulting blurring of academic with commercial objectives and a changing balance of incentives around open science.

Finally, active regions of conflict around the globe have put strain on international collaborations. As an example, in light of the ongoing conflict in the Ukraine, the CERN Council ended CERN's cooperative agreement with Russia and Belarus.[6]

17.4 What Is the Impact of Changes to the Current Science Ecosystem?

National security and economic security concerns are posing threats to the open science research ecosystem in the US.

At universities, traditionally the major locus for open global collaborative research in the US, there are increasing concerns about foreign IP theft. Inappropriate behavior on the part of some individual researchers,[7] both U.S. and foreign, has led to concerns in the US of improper dissemination of research findings.[8,9] The government has

---

[4] Center for Strategic & International Studies. Survey of Chinese Espionage in the United States Since 2000. https://www.csis.org/programs/strategic-technologies-program/survey-chinese-espionage-united-states-2000

[5] https://www.govinfo.gov/content/pkg/USCODE-2011-title35/html/USCODE-2011-title35-partII-chap18.htm

[6] https://home.cern/news/news/cern/cern-council-decides-conclude-cooperation-russia-andbelarus-2024

[7] https://www.justice.gov/usao-ma/pr/former-harvard-university-professor-sentenced-lying-about-his-affiliation-wuhan

[8] https://www.nytimes.com/2019/11/04/health/china-nih-scientists.html

[9] https://www.chemistryworld.com/news/theft-of-universities-secrets-fuels-us-crackdown-on-chinese-talent-programmes/4011014.article

responded with refinements to long standing policies around openness in government funded R&D (such as NSDD 189[10]) with additional requirements for participation of non-US citizens in government funded R&D at universities. An example is NSPM-33[11] issued by President Trump at the end of his first term and implemented under President Biden, that increases disclosure requirements for researchers to receive federal funding and requires increased vetting of foreign students and researchers participating on projects. There is also increasing use of a somewhat ill-defined category of 'controlled unclassified' to protect information that is deemed sensitive.

The discussions in the US are challenging. Those who advocate for free exchange of ideas may downplay or dismiss threats of unfair practices in the open international research environment. Those focused on threats posed by foreign adversaries may be underestimating the negative impact of restrictions on research collaborations. The outcome of these discussions has the potential to lead to more isolationist attitudes and policies going forward.

17.5 The Path Forward

As we contemplate the path forward, we first must acknowledge that we will not be going back to the ways that international collaborations were handled in previous decades. While we can hope that the conflict regions in Ukraine and the Middle East will find resolution soon, national economic and national security concerns will not go away and will continue to influence our open research ecosystem. Going forward, the benefits of international engagement and collaboration (such as better science, science at scale, national interest and science diplomacy,) will need to be delicately balanced against risks such as the loss of competitive advantage in the commercialization and potential development of military technologies. This author believes that in many cases in the future, as in the past, the benefits of international engagement will outweigh the risks, especially in early-stage basic research. However, there are cases where we will need new risk mitigation strategies as the basic research moves to technology development and deployment. And while our primary focus should be first and foremost on fostering innovation and economic development globally, we must also continue to work to mitigate potential threats to national security.

There are three elements to a framework to mitigate potential threats to national security from our open research ecosystem. All of these exist in some form currently, but we need to improve them to accommodate the current geopolitical climate.

[10] https://www.acq.osd.mil/dpap/dars/pgi/docs/National_Security_Decision_Directive_189.pdf

[11] https://trumpwhitehouse.archives.gov/presidential-actions/presidential-memorandum-united-states-government-supported-research-development-national-security-policy/

17.5.1 Articulate What we Are Trying to Protect

We need a clearer articulation of what it is we are trying to protect in our research environment. In the US, we have a long history with the security classification of technology development for items with military value and the system works well. For research that does not meet the threshold for classification, how conservative do we want to be? And why? In 1985 the Regan administration articulated the policy that "to the maximum extent possible, the products of fundamental research remain unrestricted".[12] Recent federal guidance focuses more on taking steps to protect intellectual capital and discouraging 'research misappropriation'.[13] The balance between protectionism and openness will be helped by more clarity around what we are trying to protect. As one Department of Energy national laboratory director has stated: "Protect what we need to protect, but don't do so in a way that undermines our ability to develop S&T worth protecting.[14]" It is imperative that scientists engage with policy makers to reach the best outcome.

17.5.2 Exercise Caution

Research theft and misappropriation by foreign state actors occurs, and it threatens the open research environment. It may be necessary to become more cautious about whom we collaborate with and on what topics [5]. It will help to have greater clarity around expectations for collaboration participation that model what is often done for large international collaborations. And US universities may need to pull back from some of their global research engagements. However, we should not pull back from engaging students from around the globe in our university research enterprise.

17.5.3 Develop New Strategies

We need to develop more strategies for positive international collaboration to address global challenges with low dual use potential and tremendous negative economic impact if the problems are not solved. Examples include climate change, food security and global health. Global collaboration is required to solve these problems, and working together to overcome shared challenges is a way to build trust. We should take a long view on the economic benefits of the collaborations and focus less on tactical near-term economic risks of technology sharing and more on the strategic long term economic benefits of getting to solutions.

---

[12] https://www.acq.osd.mil/dpap/dars/pgi/docs/National_Security_Decision_Directive_189.pdf

[13] https://trumpwhitehouse.archives.gov/presidential-actions/presidential-memorandum-united-states-government-supported-research-development-national-security-policy/

[14] Dr. Thomas Mason, Director, Los Alamos National Laboratory

17.6 Conclusions

International science collaboration and science diplomacy will be ever more important in the future and as a result, we, scientists, diplomats and policy makers, need to continue to invest to preserve an open and collaborative global research ecosystem. This author believes that the two most important components of that open global research ecosystem for the US are and will continue to be that we continue to welcome scholars from around the globe to study, learn and contribute, and that, to the maximum extent possible, we keep the products of fundamental research open and unrestricted.

It is a pleasure to acknowledge the organizers and participants of the XXIII Amaldi Conference for stimulating discussions and thoughtful presentations. I would also like to thank colleagues Norbert Holtkamp, Kam Moler, Micah Lowenthal, Saurabh Anand and Raymond Jeanloz, all of whom, contributed to the framing of this paper. The opinions expressed and any factual inaccuracies are mine alone.

References

1. T. Flink, N. Ruffin, The current state of the art of science diplomacy, in *Handbook on Science and Public Policy*, (Edward Elgar Publishing, Cheltenham, UK and Northampton, MA, 2019), pp. 104–122
2. V. Turekian, The evolution of science diplomacy. Global Policy **9**, 5–7 (2018). https://doi.org/10.1111/1758-5899.12622
3. P.D. Gluckman et al., Science diplomacy: A pragmatic perspective from the inside. Science & Diplomacy **6**(4), 1–13 (2017) https://www.sciencediplomacy.org/sites/default/files/pragmatic_perspective_science_advice_dec2017_1.pdf
4. H. Smith, T. Neff, Who helped turn Soviet nukes into electricity, dies at 80, in *Washington Post*, (2024) https://www.washingtonpost.com/obituaries/2024/07/23/thomas-neff-dead/
5. Holtkamp N, Mason T, Puglisi A (2024) How to Create, and Sustain, R&D Leadership. https://www.hoover.org/research/how-create-and-sustain-rd-leadership

Chapter 18
Nuclear Risks Today

Paolo Cotta-Ramusino

18.1 Concluding Remarks

Today, we face serious conflicts involving states that possess nuclear weapons.

The Russian—Ukrainian conflict began with the Russian invasion of the Crimea and, most importantly, with the full-scale invasion of Ukraine on 22 February 2022. There were, of course, prior tensions between Russia and Ukraine. In February 2014, the pro-Russian Ukrainian President Yanukovych was forced to leave Ukraine and was replaced by Poroshenko, followed later by Zelensky. According to UN data, approximately 14,200 people were killed in Ukraine before 2022 during internal fighting between pro-Russian separatist forces and Ukrainian troops in the Donetsk and Luhansk regions, including 3400 civilians.

It is also important to recall that in 1990–1991 there was an explicit agreement to limit NATO's expansion to the only territory of former East Germany. This agreement was widely violated: in 1999, Poland, Hungary, and the Czech Republic joined NATO, followed by Bulgaria, Estonia, Latvia, Lithuania, Romania, Slovakia, and Slovenia before 2004. Later, Albania, Croatia, Montenegro, and North Macedonia also joined before 2020, and after Russia's 2022 invasion of Ukraine, Finland and Sweden became members. NATO's expansion contributed significantly to Russia's sense of isolation and external pressure. It should be emphasized, however, that these historical factors do not justify Russia's invasion, which remains a clear violation of international law. The purpose of this discussion is to better understand Russia's motivations in order to explore possible pathways to peace.

The number of casualties in the war following February 2022 is highly uncertain. Civilian deaths are estimated at around 15,000, with more than 34,000 civilians injured, while total military casualties—both dead and wounded—may exceed one million.

P. Cotta-Ramusino (✉)
Department of Physics, University of Milan, Milan, Italy
e-mail: Paolo.Cotta@mi.infn.it

P. Cotta-Ramusino et al. (eds.), *Nuclear Weapon Risks in a Problematic Time*, Springer Proceedings in Physics 358, https://doi.org/10.1007/978-3-032-29873-7_18

The nuclear risks associated with the Russian-Ukrainian war stem from the involvement of nuclear-armed states on both sides—Russia and NATO- which have been supporting the conflict for well over three years. Attempts to negotiate a ceasefire have so far failed. Ukraine is *de facto* permitted to use weapons supplied by the Western alliance to target sites inside Russian territory, including military installations, while Russia continues to strike Ukrainian territory, causing numerous civilian casualties. Recent Ukrainian attacks on Russian nuclear-capable bombers and Russia's retaliatory bombings in Ukraine have exacerbated the situation, leading to increasingly severe consequences for civilian life. Then there is the problem of (allegedly) Russian drones and of missiles launched against Western countries. Several drones and missiles have also been launched into Russian territory, particularly in the Belgorod region.

Arms control agreements are also under severe strain. Most direct inspections and exchange of information regarding the number and deployment of nuclear weapons, held by the two major powers are no longer taking place. There has also been discussion in Russia about the possible use of nuclear weapons if Russian territory comes under a serious attack. Moreover, there have been recent suggestions that Russia may resume nuclear testing. The only remaining arms control agreement is the New START Treaty, which limits U.S. and Russian nuclear forces to 1550 deployed warheads and 700 deployed strategic launchers on each side. This treaty is set to expire on 5 February 2026. Russia recently declared that it is prepared to continue observing the "central quantitative limits" for one additional year, provided that the US agrees to do the same.

In conclusion, despite the significant efforts made over the past decades to reduce the total number of nuclear weapons in the U.S. and Russian arsenals—from about 70,000 in the 1970s to fewer than 11,000 today –nuclear risks are arguably higher now than in the recent past. The current U.S. administration may attempt to mitigate nuclear risks associated with the Russian-Ukrainian war, but it remains to be seen how the situation will evolve.

18.2 The Conflict in the Middle East after 7 October 2023

This conflict is also highly relevant from the standpoint of nuclear risks, since Israel possesses nuclear weapons and Iran could acquire them in a relatively short period, potentially abandoning all remaining aspects of the former JCPOA agreement. Israel, with U.S. support, has reportedly attacked certain nuclear facilities in Iran. The conflict in Middle East has expanded dramatically. It began as a confrontation between Hamas and Israel but has since drawn in the entire population of Gaza, the West Bank, Iran, Lebanon and the Houthi-controlled regions, and more recently, areas in Syria and Iraq. So far, the direct exchange of fire between Israel and Iran have remained relatively contained, but it is uncertain whether this situation will continue. Additionally, regions in Iraq suspected the harboring PKK support have recently been targeted by Turkey. The number of casualties and deaths across the region is staggering. Approximately 70,000 people have been killed in Gaza—most of them civilians, including many women and children. Some people in Gaza have reportedly died from starvation. Many UN personnel, including doctors and nurses assisting wounded civilians, have also been also killed. Around 3500 people have recently been killed in Lebanon. In Israel, about 1200–1300 people were killed on 7 October. More than 1000 people have also been killed in the West Bank. The destruction of buildings and civilian infrastructure has been immense, particularly in Gaza but also in the West Bank and Lebanon. Although a ceasefire in Gaza has now been declared, even if it is fully implemented, the enormous challenges of reconstructing the devastated Gaza Strip and, above all, establishing a viable Palestinian state will still need to be addressed.

In any case, from the standpoint of nuclear risks, as mentioned earlier, the relationship between Israel and Iran remains one of the key issues to consider. Naturally, this relationship is deeply influenced by the broader regional conflicts. There have been Israeli attacks, as well as limited American strikes, on Iranian nuclear facilities. These actions have not deterred Iran from pursuing nuclear enrichment—on the contrary, they have had the opposite effect. The ability to develop nuclear enrichment has become a matter of national pride for Iran and a symbol of capacity to resist external pressure. In the past, there was some discussion in the U.S. about reaching an agreement with Iran, and it is still possible that such an agreement might be developed in the future—though this remains to be seen. One of the central issues is whether Iran should be allowed to pursue uranium enrichment for the declared purpose of producing nuclear energy for peaceful uses. In any case, such enrichment should be firmly monitored by the International Atomic Energy Agency (IAEA). It is well known that enriching uranium to 3.5% represents roughly halfway toward enrichment to 90%. However, the Non-Proliferation Treaty (NPT) explicitly states that its member states have the right to develop civilian nuclear energy. Let us hope that a possible solution to this complex issue will be found in the near future. Unfortunately, the European Council has recently agreed to reimpose a number of restrictive measures related to Iran's nuclear activities, which had previously been suspended with the entry into force of the Joint Comprehensive Plan of Action (JCPOA)—the so-called "Iran nuclear deal"—through the snapback mechanism.

18.3 The India—Pakistan Antagonism and Nuclear Weapons

Both India and Pakistan possess nuclear weapons, estimated roughly 170 each. In recent years, public attention has somewhat waned regarding the nuclear risks associated with these two countries. However, following the terrorist attack in Kashmir, tensions between India and Pakistan escalated dramatically. A substantial number of nuclear weapons remain in the hands of these two rival nations, and the possibility of a direct conflict between them, unfortunately, cannot be excluded. Pakistan has never adopted a *no first use* policy, while India, for some time, claimed that it would never be the first to use nuclear weapons. However, this Indian stance is now reportedly under reconsideration. Meanwhile, Pakistan's internal situation remains highly fragile and unstable.

18.4 China and the Taiwan Issue

China has become the third-largest nuclear power, reportedly possessing around 600 nuclear weapons—a number that appears to be steadily increasing. Although China maintains a formal *no first use* policy, it is currently the only country expanding its nuclear arsenal. Beijing's pressure on Taiwan for "reunification" has intensified compared to the past. While the possibility of Chinese military intervention in Taiwan has attacked some attention, it does not yet appear to be among the world's most urgent international concerns. This is likely because other major conflicts involving nuclear-armed states are currently dominating the global agenda. If China were to exert stronger or even direct military pressure on Taiwan, it remains uncertain how countries such as the United States—which provided significant support to Taiwan—would respond. Major powers like Russia, which heavily depend on their alliance with China, would most likely refrain from exerting any serious pressure on Beijing. Overall, the situation could become extremely dangerous.

18.5 The Situation in the Korean Peninsula

It is well known that North Korea (the DPKR) possesses nuclear weapons—around 30 or more. North Korea has now openly supported Russia in the war against Ukraine, reportedly sending troops to assist Russian military operations. This has strengthened the Russia-DPRK alliance. In the event of a conflict involving North Korea, South Korea or other countries, the implications for nuclear risks would be highly significant.

Concerning nuclear risks, it should also be pointed out:

(a) When the global situation becomes more critical and unstable, some countries may see such conditions as opportunities to advance their own specific interests. Consequently, conflicts may trigger the development of new conflicts, further increasing nuclear risks or even promoting nuclear proliferation.
(b) In today's world, political leaders increasingly view popular support as a crucial instrument for maintaining power. As a result, the consequences of their actions for global stability risk becoming less relevant than the domestic approval they can secure—or believe they can secure. This applies, for example, to Netanyahu, Putin, and the U.S. President. By contrast, during the Cuban Missile Crisis,

leaders such as Khrushchev and Kennedy focused more on containing the global nuclear risks than on preserving their own political positions.

(c) It is very important that the scientific communities cooperate internationally to advise political leaders about the catastrophic consequences of nuclear explosions and the urgent need to minimize the risk of a nuclear war.

(d) Let us once again emphasize a crucial point: when wars occur between states that possess nuclear weapons, the probability that nuclear arms might eventually be used must be taken with the utmost seriousness. Every possible effort should therefore be made to bring those states to a ceasefire and to negotiate a workable, lasting agreement.

FSC
www.fsc.org
MIX
Papier | Fördert gute Waldnutzung
FSC® C083411

Zeitfracht Medien GmbH
Ferdinand-Jühlke-Straße 7
99095 Erfurt, Deutschland
produktsicherheit@kolibri360.de